Daniel Hack Tuke

The Insane in the United States and Canada

Daniel Hack Tuke

The Insane in the United States and Canada

ISBN/EAN: 9783337375607

Printed in Europe, USA, Canada, Australia, Japan

Cover: Foto ©berggeist007 / pixelio.de

More available books at **www.hansebooks.com**

THE INSANE

IN THE

UNITED STATES

AND

CANADA.

BY

D. HACK TUKE, M.D., LL.D.,

FELLOW OF THE ROYAL COLLEGE OF PHYSICIANS, LONDON.
CO-EDITOR OF "THE JOURNAL OF MENTAL SCIENCE."

LONDON:
H. K. LEWIS, 136, GOWER STREET, W.C.
1885.

To
THE MANY FRIENDS
ON THE AMERICAN CONTINENT
WHO AIDED HIM IN CARRYING OUT THE PURPOSE OF HIS VISIT TO THE
UNITED STATES AND CANADA,
THIS SMALL CONTRIBUTION
TO A GREAT OBJECT
IS GRATEFULLY DEDICATED
BY
THE AUTHOR.

CONTENTS.

CHAPTER I.

EARLY LUNACY PRACTICE IN AMERICA.
BENJAMIN RUSH, M.D.

Sketch of Rush's Life—His Medical Practice—His Death. Medical Works—Observations on Old Age—On the Duties of a Physician—Rush in Relation to the Insane—Moral Treatment—Stratagem—Erect Posture—Terror and Pain—The Gyrator—Prolonged Fasting—Importance of Labour—Enlightened Views—Dipsomaniacs—Medical Treatment-Baths, etc.—Blood-letting—Defence of it by Rush—Opposite Practice at York Retreat—Bleeding in Mania—Directions for Bleeding—Kirkbride and Earle on Bleeding—Pargeter—Crowther—Haslam—Monro—Esquirol—Sir Alexander Morison—Burrows—Prichard—Bell—Todd—Brigham—White—Ray—Curwen—Rush on Moral Insanity—Case of Servin—Genius and Moral Depravity—Doubtful Irresponsibility—Rush, the American Fothergill—Reflections on the Death of Rush.

CHAPTER II.

PROVISION FOR THE INSANE IN THE UNITED STATES, 1752 TO 1876.

Pennsylvania Leads the Way—Williamsburg, Virginia—Frankford Asylum—Miss Dix—Loathsome Crazy-houses—Noble Appeal by Miss Dix—New Hampshire—Dr. Bell—Dr. Ray—Dr. Brigham—Dr. Jarvis—Dr. Woodward—Dr. Howe—Dr. Wilbur—Memorial of Pennsylvania Medical Society—Board of Public Charities—Petition from Philadelphia—Physicians and Boards of Charity—Increase of Asylums—Admitted Duty of the States—The *Lancet* and Dr. Bucknill—Hospitals for the Insane.

Contents.

CHAPTER III.

PRESENT CONDITION OF THE INSANE IN THE UNITED STATES.

Number in Restraint in 1880—The Covered or Crib-Bed—Complaint of American Alienists—Seclusion—Padded Rooms—Suicidal Cases—Medical Treatment—Female Physicians—Recoveries and Deaths—Type of Mental Disorder—Employment of Patients—Less in America than England—Lunacy Legislation—Admission Papers—Practice in Different States—Insane Thursday—Jury-Trial—Its Disadvantages—Massachusetts Law—Inspection—State Board of Massachusetts—Pennsylvania and New York—Provision for Chronic Insane—Distribution of Insane—Dr. Chapin and "Willard"—Kankakee—Mr. Wines—Dr. Dewey—Detached Buildings at Washington—Middletown, Norristown, etc.—The System in Wisconsin—Dane County Asylum—Dodge County Asylum—Rock County Asylum—Dangers of this System—Conditions of Success—Relative Merits of American and English Asylums—Salaries—Black List—State Provision for all Classes—Classification—Segregation—Lady-Physicians—Liberal Diet—Ward Decorations—Restraint—Future of Asylums Encouraging—Difficulties from Immigration—Foreign Insane Inheritance—Correction by Mr. Wines.

CHAPTER IV.

PRINCIPAL ASYLUMS VISITED.

Visit to the Concord Asylum—New Hampshire Almshouses—Their Bad Condition—State Asylum Necessary—Vermont Hospital for the Insane—Location of Insane in Vermont—Massachusetts—Northampton Hospital—New England Psychological Society—McLean Asylum—Social Science, Saratoga—Utica Asylum—Form of Certificate—Danvers Asylum—Worcester Chronic Asylum—State Asylum, Worcester—Butler Hospital for Insane—State Asylum, Rhode Island—Middletown Hospital—Hartford Retreat—New York City Asylum—Emigrant Hospital for Insane—Bloomingdale Asylum—New Jersey State Asylum—Hudson County Asylum—Penn. Hospital for Insane—Norristown Asylum—The Frankford Asylum—Harrisburg Asylum—Government Hospital for Insane—Kankakee—Milwaukee Asylum—Wisconsin State Hospital—Maryland State Hospital—Mount Hope Asylum—Summary.

Contents.

CHAPTER V.

THE INSANE IN CANADA.

Longue Pointe Asylum — Beauport Asylum — Important Resolutions adopted by Medical Men at Montreal—Location of Insane in Ontario —Recoveries, Deaths, and Cost—Employment — Lunacy Law of Canada—License for Private Asylums—Ontario and Quebec Contrasted—Toronto Asylum—Forty Years Ago—London Asylum—Hamilton Asylum — Kingston Asylum — Kingston Penitentiary — Asylum of New Brunswick—Asylum of Nova Scotia—Prince Edward's Island—Manitoba—Newfoundland.

APPENDICES.

(A.) TABLE I.—General Statement of Insane in Hospitals and Lunatic Asylums in the United States and British Provinces, Sept. 30, 1883.

(B.) TABLE II.—The same, in detail, showing the name of every Institution, the number of patients, and the names of the Superintendents.

(C.) TABLE III.—Showing Total Number of Insane in each of the United States, and the proportion to the population.

(D.) TABLE IV.—Showing amount of Restraint and Seclusion in twenty Hospitals for the Insane in 1882.

(E.) TABLE V.—Showing the number of Idiots in each State in 1880, and their Location.

(F.) TABLE VI.—Showing the number of Idiots in Training Schools for the Feeble-minded in 1880.

(G.)—Memoranda on average cost of construction of Asylums, the total annual expenditure on the Insane, and the average annual cost per patient.

PREFACE.

In the following pages the writer has endeavoured to record briefly, but he hopes correctly, the impressions received in regard to the condition of the insane in the United States and in Canada during a recent tour in these countries.

In the first chapter he has sought to describe the early treatment of the insane, medically and morally, in the States, at a period in which Dr. Rush formed the central figure on the stage of medical practice.

The substance of the second chapter was written in 1876, and contains a sketch of the provision made for the care of the insane up to that time, which had the advantage of the late Dr. Ray's revision. It must be read from the standpoint of the period at which it was written. The hospitals for the insane opened between that and the present year are added in order to complete the series.

The continuation of the narrative is pursued in the account given of the actual state of the American asylums at the present time, so far as the writer's personal knowledge enables him to describe it.

For the notice of individual institutions which follows, he must claim the indulgence of those who preside over them. It has been found difficult to decide how much to give of the copious information obtained from superintendents and others, to all of whom he must express his grateful sense of obligation. The attempt has been made to bring into relief, the points which chiefly struck the visitor, rather than to burden his pages by transferring the contents of his note-book to them. He is only too well aware that he must have fallen into some errors, but he asks his American friends, and he believes he will not ask them in vain, to attribute these to any other cause than intentional mis-statement. If he has been so fortunate as to avoid serious errors, he does not doubt that he has omitted to notice some, probably many, features in the asylums visited, worthy of mention and deserving of praise.

It is a cause of regret that the condition in which he found certain asylums in Canada left him no alternative, consistent with truth, but to paint them in the colours in which they presented themselves to his view, while gladly admitting that there were bright lights as well as dark shades in the picture. He does not think there is anything in the attacks made by some of the Canadian newspapers (notably the *Minerve* and the *Étendard*) which merits serious reply. To the question asked by one of these journals, "Franchement! est-il possible, surtout à un scientist, de se montrer plus sot, plus fourbe, plus insolent?" he is content to allow the *Montreal Daily Witness* to reply: "His cautious and kindly report was duly published in a medical journal. As it was evident that such a statement as Dr. Tuke's would easily withstand the shot and shell of gallant and devout adulations which had been stored within its walls, and which were hurriedly poured forth for purposes of defence, nothing remained but abuse; and the ignorance, slavishness, and inhumanity with which this has been vented by the paper named, only make that paper look foolish."

It is to be hoped that a Bill which has been introduced into the Canadian Parliament, "respecting Lunatic Asylums in the Province of Quebec," will, if passed, effect some good, but abuses will never be satisfactorily provided against so long as the present system is permitted to continue in force. Radical remedies will alone suffice to remove radical evils.

If the writer's blame or praise of institutions on the American Continent produces any good results in the interests of the insane, his object will have been abundantly attained.

D. H. TUKE.

Lyndon Lodge, Hanwell, W.
August 12, 1885.

THE INSANE IN THE UNITED STATES AND CANADA.

CHAPTER I.

EARLY LUNACY PRACTICE IN AMERICA.
BENJAMIN RUSH, M.D.

HAVING for many years held this celebrated American physician in the highest esteem, and taken occasion more than once to express my appreciation of his admirable writings in relation to mental affections and also the influence of the mind upon the body, I made a point of inquiring when in Philadelphia last September whether a portrait of him existed. I ascertained at the General Hospital that they possessed an oil painting, by Sully, which pourtrayed the Philadelphia physician. I was courteously shown it, and although it did not exactly correspond with the mental image I had formed of the man, it had every appearance of being a life-like portrait. I could not ascertain that it had been copied, but the physician who showed me the painting promised to obtain a copy for me if it existed. He, however, failed.

Dr. Richardson has told the readers of his "Asclepiad" for January last how, on meeting Dr. Weir Mitchell in London last year, he conversed with him about Dr. Benjamin Rush, and how Dr. Mitchell promised to supply him with a portrait. He succeeded, and an excellent impression executed by the Autotype Company, London, appeared in the January number of the above-mentioned journal, which quarter by quarter bears witness to the extraordinary mental energy and ability of the editor. To my request for permission to make use of it, Dr. Richardson generously responded, and I am sure my readers will thank him for granting me this favour.

I proceed to give a brief enumeration of the principal events of Dr. Rush's useful and active life.*

He was born near Philadelphia, December 24, 1745. He was educated at the academy of an uncle, the Rev. Samuel Finley, in Nottingham, Maryland, where he resided eight years. He graduated as B.A. at Princetown, New Jersey, in 1760; he then studied medicine under two American physicians, and subsequently, from 1766 to 1768, at Edinburgh, where he obtained his degree of M.D. After spending a week in London he returned to Philadelphia, and was elected Professor of Chemistry in the College there. He was 46 years of age before he filled the chair of medicine in the University of Pennsylvania, into which the College merged. His name is found among the signatories to the Declaration of Independence in 1776, when he was a member of Congress. In the following year he was appointed physician to the Military Department of the hospital. That he was not so absorbed in his profession as to be unable to devote some time to public matters is shown by his acting as a member of the Convention for adopting the Constitution of the United States, and by fulfilling the duties of treasurer to the United States Mint during the last fourteen years of his life. He was president of the Society for the Abolition of Slavery, and he was connected with a number of charitable, religious, and literary societies. He wrote strongly against the use of tobacco on account of its deleterious effect upon the health and morals. He was among the first to advocate temperance. He was the patron of education, and devised the establishment of free schools.

* The particulars which follow are mostly given on the authority of the "American Biographical Dictionary," by Allen, 1857.

It is stated that on commencing practice in Philadelphia Dr. Rush found the system of Boerhaave paramount in medicine, and that he publicly taught that of Cullen in opposition to it. After a while, however, he threw Cullen entirely aside, and refusing to be shackled by any one master, became an original observer and teacher. He simplified the interminable list of diseases which the Edinburgh Professor had invented, and with which he had bewildered the medical student. To Rush, names were of little importance; the reality—the essential nature—of the disease was what he strove to grasp and teach. While the system of Cullen demanded theoretical nosologies and burdened the memory, that of Rush appealed to clinical facts and common sense. Dr. Thacker, who knew both systems from practical experience, feelingly observes that it is no exaggeration to say that in three years, the one, the American, prepared the student for the art of healing, more thoroughly than the other, the Scotch, did in five. Rush so mastered the indications of the pulse that instead of being "fallacissima res," it was to his subtle and practised touch, the "nosometer of the system."

That Rush bled and purged and used calomel (which he designated the Samson of the Materia Medica) to a degree which we have come to regard as deadly, cannot be denied. Whether, like the Hebrew hero, mercury has, as its enemies allege, slain its thousands, will always be a matter of debate; but looking at the marvellous success and popularity of the Philadelphian physician, one finds it hard to believe that this, or any of the heroic remedies which he employed, had the same effect upon the human system that they have at the present day. Rush always maintained that a treatment which even then might be granted to be unsafe in Europe was required in America. At any rate, when the yellow fever raged in Philadelphia in 1793, and every other remedy (including bark, wine and cold baths), failed, his favourite remedies, which in this disease he tried with hesitation and fear, succeeded.

Even when the pulse was scarcely perceptible, purgation by calomel restored the patient. When the tide thus turned he wrote in his note-book, "Thank God, out of one hundred patients whom I have visited or prescribed for this day, I have lost none!" He himself nearly fell a victim to the incessant labour which his success entailed upon him, but after free bleeding and purging he recovered. The mouth of criticism is stopped.

For close upon half a century Dr. Rush gave the results of his experience and meditations to the world. His style, however, bears no marks of haste, notwithstanding his constant occupation. He held that our ideas, whether original or acquired, produce a mental plethora, which depletion by the pen or tongue can alone relieve. The pen, therefore, was to him in his literary work what the lancet was in his practice. His admirable devotion to his profession is expressed with his usual felicity in a letter, thus—"Medicine is my wife; science is my mistress; books my companions; my study is my grave. There I lie buried, 'the world forgetting, by the world forgot.'" Medicine was not, however, to be his only wife. Forgetting or disregarding his philosophic resolutions, at the age of 32 he married Julia, daughter of Judge Stockton, and had a large family. He held that no female should marry before 16, and no male before 21, and that the longer they put off "perpetrating matrimony" the better, so long as the former do not delay beyond 24 and the latter beyond 30.

Rush died of pleurisy after a few days' illness on the 19th of April, 1813, aged 67. He had been bled, but the scythe of death vanquished the lancet of the physician.

The principal works of Dr. Rush are contained in seven volumes. Of these the first four consist of Medical Inquiries and Observations. The first edition was published from 1789-98, and appeared in five volumes. The second edition was published in four volumes in 1804.

The fifth volume, entitled " Medical Inquiries and

Observations upon the Diseases of the Mind," appeared first in 1812, and passed through several editions, a fifth being published in 1835.

A sixth volume contains sixteen introductory lectures to Courses on Medicine, with two lectures on the Pleasures of the Senses and of the Mind, and bears the date of 1811.

The seventh volume consists of Essays, "Literary, Moral, and Philosophical." The first edition was published in 1798.

In 1809, Dr. Rush edited "Sydenham's Works" as well as those of other medical authors, and to these works he contributed valuable notes.

I should have stated that his Latin Thesis, written when he graduated in 1768, was entitled "Dissertatio Physica de Coctione Ciborum in Ventriculo."

In his "Medical Inquiries and Observations," the chapter entitled "An Account of the State of the Body and Mind in Old Age, with Observations on its Diseases and Remedies," possesses lasting value, and exhibits that acute observation of the small as well as the large circumstances of life in health and disease which so strikingly characterizes this physician. In it he observes that he has not met with a single instance of an octogenarian whose descendants were not also long-lived. He only knew one instance of an unmarried person living beyond the age of eighty. He records the case of one woman in England in the hundredth year of her age who had borne a child at sixty, and frequently suckled two of her children at the same time. Most of her life was spent over a washing-tub. He pointed out the influence of emigration in the decline of life in imparting fresh vigour and apparently prolonging life for many years. He had only met with one person beyond eighty who had ever been affected by a disease of the stomach, and he was struck with the circumstance that a gentleman who died in the 109th year of his age had never vomited in his life, although he had been at sea several years when young. This man's mother lived to 91, and his grandfather to 101. It is stated that he had a great dislike to spirits, and that his constant drinks were water, beer, and cider. In reference to the memory of old persons, after making the common observation in regard to recent events and those occurring long ago, Rush records the case of a German woman who had learned to speak English after she was forty and had forgotten it entirely after the age of eighty, while she spoke her native tongue as fluently as ever. He comments on the increase of appetite in old people, and the recurrence of the desire for articles enjoyed in childhood, especially sweet substances, as cake, &c. Again, the circumstance did not escape his observation that the aged resemble children in a tendency to detail immediately to others all they see and hear. Many other striking observations are made in this essay, but I have adduced enough for my present purpose, which is to indicate to the reader unacquainted with the writings of Dr. Rush the acuteness of observation and the many-sidedness which characterize them. The address on the Duties of a Physician, delivered in 1789, is replete with practical advice. In it he says: "Permit me to recommend to you the study of the anatomy of the mind, commonly called metaphysics. The reciprocal influence of the body and mind upon each other can only be ascertained by an accurate knowledge of the faculties of the mind and of their various modes of combination and action. It is the duty of physicians to assert their prerogative, and to rescue the mental science from the usurpations of schoolmen and divines. It can only be perfected by the aid and discoveries of medicine." He particularly recommends in the way of medicinal treatment the indigenous medicines of America. "Who knows," he says, "but it may be reserved for America to furnish the world, from her productions, with cures for some of those diseases which now elude the power of medicine? Who knows but that, at the foot of the Alleghany Mountain, there blooms a flower that is an infallible cure for the epilepsy? Perhaps on the Monongahela or the Potomac, there may grow a root that shall supply by its tonic powers the invigorating effects of the savage or military life in the cure of consumptions."

I shall now endeavour to determine the position which ought to be accorded to Rush in relation to the insane and their treatment.

In his writings, so far as they bear directly upon insanity, he not only dwells upon the theoretical aspect of the subject, but supports his views by a frequent reference to very interesting cases.

Of the condition of the insane, or their claims on the State, he makes little mention. I am not aware that there is anything which shows that he was struck with the inadequate provision made for them or their miserable state at that time in gaols and almshouses, and even in the cells of the basement of the Pennsylvania Hospital in Philadelphia. These receptacles survive as relics of the past to indicate the wretched provision made for maniacs in those days. Coercion in some form of personal restraint was regarded by everyone as absolutely necessary; and the views entertained by Rush, and the practice he pursued, differed little in this respect from those then in vogue.* What I should claim for him would be that he distinctly recognised the corporeal nature of insanity; that to his students and in his writings he taught that it is a disease that must be submitted to medical as well as moral treatment; and, further, that he gave to the profession and to the world an able exposition of the forms of mental disorder.

When, in 1812, he wrote his classic work on the Mind the experiment of the treatment of the insane by very different methods from those in fashion had been in opera-

---

* As will be seen, he objected to chains and stripes.

tion at the York Retreat for some years; had, in fact, ceased to be an experiment.

A few passages from his work will indicate his opinions on the moral treatment of the insane. In mania he recommended the following modes of coercion when milder means had been employed without success :—First, the strait-waistcoat or a chair called the "tranquillizer," which is another name for the well-known restraint chair; secondly, privation of the patient's customary food; thirdly, pouring cold water into the coat-sleeves so that it might descend down the trunk and body generally; fourthly, the shower-bath continued for 15 or 20 minutes, which one would wish to believe a misprint for seconds. Dr. Rush adds that "if all these modes of punishment fail of their intended effects, it will be proper to resort to the fear of death." He gives an edifying example of the success of this *dernier ressort*. A certain Sarah T. disturbed the whole hospital by her loud vociferations. Light punishments and threats had failed to put a stop to them. The gentleman in charge of the case, Mr. Higgins, at last went to her cell and conducted her, loudly vociferating, to a large bathing-tub, in which he placed her. "Now," said he, "prepare for death. I will give you time enough to say your prayers, after which I intend to drown you by plunging your head under this water." The patient immediately uttered a prayer, such, we are told, as became a dying person; then Mr. Higgins, satisfied with this sign of penitence, extorted from her a promise of amendment. We are assured that from that time no vociferations or maledictions proceeded from the cell of Sarah T. I think I ought to add that within the memory of living man a similar experiment was tried in St. Luke's Hospital upon a male patient with the plunge-bath, and with a like satisfactory result! Dr. Rush's commentary must be added : "By the proper application of these mild and terrifying modes of punishment, chains will seldom and the whip never be required to govern mad people. I except only from the use of the latter those cases in which a sudden and unprovoked assault on their physicians or keepers may render a stroke or two of the whip* or of the hand a necessary measure of self-defence." We are sure that Dr. Rush would have exercised any of these coercive measures with the greatest reluctance, and, when he deemed them necessary, with the greatest possible gentleness; but the observations which he makes throw a curious light upon the ideas and practices of his day, from which even so enlightened and admirable a man as Dr. Rush could not entirely free himself. In passing, it may be remarked that Rush felt much confidence in relieving patients of their delusions by stratagem, for he says that "cures of patients who suppose themselves to be glass may easily be performed by pulling a chair upon which they are about to sit from under them, and afterwards showing them a large collection of pieces of glass as the fragments of their bodies."

There is a curious passage in Rush's work in which he speaks of treating refractory patients by forcing them to retain *the erect position of the body*. To many it will sound barbarous, but it was certainly not intended to be so. He says, "There is a method of taming refractory horses in England by first impounding them, and then keeping them from lying down or sleeping, by thrusting sharp pointed nails into their bodies for two or three days and nights. The same advantages, I have no doubt, might be derived from keeping madmen in a standing posture, and awake, for four-and-twenty hours, but by different and more lenient means. Besides producing several of the effects of the tranquillizing chair, it would tend to reduce excitement by the expenditure of excitability from the constant exertion of the muscles that support the body. The debility thus induced in these muscles would attract morbid excitement from the brain and thereby relieve the disease.

---

* It is clear from this that the whip was considered at that time an essential part of the furniture of an asylum-ward.

That benefit would be derived from preventing sleep, I infer from its salutary effects in preventing delirium, and from delirium being always increased by it in fevers of great morbid excitement."

Dr. Rush recognised the influence of *terror*. "It should be employed in the cure of madness." He was struck with its effect upon a lady who was deranged, and for whom he had advised riding. One day the horse ran away with her, but was stopped in its course by a gate. When the attendants came up to her they found her restored to reason, and she had no return of her malady. He relates the case of a madman who was suddenly cured by the breaking of the rope to which he was attached for the purpose of bathing him in a well. The poor fellow was nearly drowned, but was cured of his insanity. The comment Rush makes upon this recovery is, that the new action induced in the brain by the powerful stimulant of terror brought about the cure.

When Dr. Rush refers to instances in which flagellation appeared to produce a salutary effect upon the insane, he is careful to say that "in mentioning the cures performed by the whip, let it not be supposed that I am recommending it in this state of madness;" and he adds, "Fear, pain, and a sense of shame may be excited in many other ways that shall not leave upon the memory of the patient the distressing recollection that he owes his recovery to such a degrading remedy." Here he was distinctly in advance of Cullen.

We cannot be surprised that Dr. Rush was fascinated, like so many others, with the famous rotatory machine of Dr. Cox. The former contrived one himself for use at the hospital, and called it the Gyrator. He also suggested a cheap contrivance, by which a patient was placed upon a board, moved at its centre upon a pivot, with his head towards one of its extremities. A rotatory motion was then to be given to the machine.

Among the remedies for madness, Dr. Rush enumerates

"great pain," and records the fact of a transient gleam of reason having been induced in several idiots in Italy by means of torture inflicted by priests. It is hardly necessary to say that the humane physician of the Pennsylvania Hospital would not have followed the examples of the Italian ecclesiastics. What he says is that "should this remedy be resorted to it should be induced by means that are not of a degrading nature, and which are calculated at the same time to excite violent passion or emotion of the mind."

It is surprising to find Rush disposed to treat some cases of insanity by prolonged *abstinence*. He suggested fasting for two or three days, and he bases this treatment upon the fact that in India wild elephants, when taken, are always tamed by depriving them of food until they discover signs of great emaciation.

But let us turn to his more enlightened views. Dr. Rush fully recognised the importance of labour. He mentions the case of an insane gentleman whom he attended, but failed to cure by his prescriptions. After his return, however, to his family in Maryland, at the time of hay-making, he was induced to take a rake into his hands and assist. He occupied himself for some time in the hay-field and recovered. Dr. Rush attributes this result to profuse perspiration. I think that it is due to the doctor to quote the following passage. "I have often wished," he writes, "and lately advised that the mad people in our hospital should be provided with the tools of a number of mechanical arts. Some of them should be laborious and employ the body chiefly; others ingenious and of a nature to exercise and divert the mind more than the body. None of them should be carried on by instruments with which it would be easy to hurt themselves or others. . . . The advantages of thus producing a current of new action, both corporeal and mental, which should continue for weeks and months, and perhaps years, could not fail of being accompanied with great advantages."

Dr. Rush observes that the inadequate means employed for ameliorating the condition of the insane leads us to lament the slow progress of humanity in their case more than in any other of the afflicted children of men. He expresses his satisfaction that the period of cruelty and insensibility to their sufferings is passing away. He refers to the "humane revolution" that had recently taken place in the receptacles for the insane in Great Britain, and he states that a similar change had occurred in the Pennsylvania Hospital. "The clanging of chains and the noise of the whip are no longer heard in their cells. They now taste of the blessings of air and light and motion in pleasant and shaded walks in summer, and in spacious entries warmed by stoves in winter." He admits that much remains to be done for their comfort and relief, and bids those whom he addresses to recollect the greatness of the object, for it is not to cure a common disease, but "to restore the disjointed or debilitated faculties of the mind to their natural order and offices, and to revive in the patient the knowledge of himself, his family, and his God."

It should be stated that, even in his day, Dr. Rush pleaded for the establishment of curative institutions for dipsomaniacs. He says, "To the account of physical remedies for drunkenness I shall add one more, and that is the establishment of a hospital in every city and town in the United States for the exclusive reception of hard drinkers. They are as much objects of public humanity and charity as mad people. . . . Let it not be said that confining such persons in a hospital would be an infringement upon personal liberty, incompatible with the freedom of our governments. . . . To prevent injustice or oppression, no person should be sent to the contemplated hospital or *sober-house* without being examined and committed by a Court, consisting of a physician and two or three magistrates, or commissioners appointed for that purpose" (p. 266). In these institutions he would not allow the use of tobacco or ardent spirits. What the substitutes for these should be we are not told. I think that he would not have forbidden beer or claret.

Let us now turn to his medical treatment. Dr. Rush recommended the application of cold water to the body, in cases of mania. He observes that the whole body should be immersed in it for several hours to prevent the reaction of the system. The same end, he adds, would be attained by pumping water on a patient for an hour or two, but this he thinks undesirable, as it might be regarded in the light of a punishment, inasmuch as it was employed in cases of drunkenness.

He recommended the cold shower-bath in succession to the warm bath, which was to be heated above the natural temperature of the body. He kept a "torpid patient" in the latter for an hour or two, and then led him, "smoking with vapour," to the shower-bath, which, he observes, gave the most powerful shock to the system. He adds, " It extorted cries and groans from persons that had been dumb for years." Rush records one case in which this mode of treatment relieved, and another in which it restored the patient. It is a sad commentary upon the miserable receptacles then in use in the Pennsylvania Hospital, that the latter patient died some time after his recovery, from the damp cell in which he was confined.

No reference to Dr. Rush would be complete without dwelling at some length upon his strong opinions and equally strong practice in regard to bleeding the insane in many forms of their disorder. It is a remarkable fact that the treatment he pursued and that adopted contemporaneously at the York Retreat were in this most important particular diametrically opposed to one another. While Rush used the lancet freely, it was almost, if not wholly, laid aside at the Retreat, nourishing food and stimulants being administered in its place. A London medical journal expressed, not only its astonishment at this daring innovation on a time-honoured practice, but its hearty approval in consequence of the satisfactory results which so manifestly followed its adoption.

But was not Rush equally successful? It is difficult, as I have said when speaking of his general practice, to believe, considering his popularity, that he was unsuccessful. Would not his own acute mind and quick observation have soon led him to the conclusion that bleeding and low diet tended to induce dementia or death, if they really led to these results in America? That there succeeded a reaction against his treatment, in his own land, it will be our purpose shortly to prove, and there are those who hold that (as has been said in regard to old-fashioned medical treatment in general) many a weeping Martha might have addressed the Doctor when he lost his patient with— "Sir, if thou hadst *not* been here, my brother had not died;" but we must not too hastily conclude that while the treatment adopted at the York Retreat was right, as no one now disputes, Rush's treatment was wrong at that time, in the country in which he practised. Those who maintain that there has been a change in the type of disease may have truth on their side, and may be correct in their opinion that insanity was no exception in those days to the general rule which, as they believe, holds good, namely, that disease at that period presented a sthenic form, and consequently required depletion. The most serious objections to this theory, when applied to England, are that depletion was thought to be necessary at Bethlem and St. Luke's, at the same time (a fact that was doubtless well known to Dr. Rush when residing in London in 1768), and that its non-necessity, nay, its absolute harmfulness, was demonstrated by the success of the opposite course of treatment pursued at York. The question thus raised is, it must be admitted, a difficult one to answer; fortunately it is one of historical interest, not practical importance, at the present moment, for all are agreed, whether in Europe or America, that mental disorders are associated with low nutrition, and do not call for venesection, however useful local depletion may occasionally be.

Fully to appreciate Rush's treatment of the insane by phlebotomy, it should be remembered that, after rejecting the prevalent notion that madness is situated in the mind itself, or in the abdominal viscera or the nerves, he held that the primary seat of insanity is in the blood-vessels of the brain, and that it depends upon the same kind of morbid and irregular action that accompanies other arterial diseases. "There is," he adds, "nothing specific in these actions. They are a part of the unity of disease, particularly of fever, of which madness is a chronic form, affecting that part of the brain which is the seat of the mind." He supports his views by many reasons, among which are—the frequency of the pulse in mania; the influence of digitalis in reducing its frequency, and at the same time mental excitement; the alternation of insanity with several diseases evidently (to him) seated in the blood-vessels, as consumption, rheumatism, &c.; the blending of the symptoms of insanity and fever; the occurrence of a form of mania attended with feeble pulse, muttering delirium, &c., recognised at that time as typhomania; the character of the blood in maniacs; the appearance of the brain on post-mortem examination; and lastly, the effects of treatment, which, according to Rush, were identical with those successfully pursued in fever. He did not restrict the seat of madness, however, to the cerebral blood-vessels, but admitted that it extended to the brain itself and the nerves. Rush fully recognised what he called the phrenitic predisposition; it consisted, he held, in a preternatural irritability of that part of the brain which is the seat of the mind and of the nerves. In consequence of this, a deranged action was more promptly communicated to the blood-vessels of the brain. The presence of this predisposition accounts for some persons being delirious in fever; its absence explains why others have disease of the brain or fever without becoming mentally deranged.

I may here mention a circumstance of interest as bearing directly upon the present subject. In the spring of 1811, one year before the publication of Rush's work on the Mind, my father addressed a letter to the governors of the New York Hospital in reply to an inquiry regarding the treatment pursued at the York Retreat. In this letter he writes:—" General bleeding and other evacuants have been found injurious at the Retreat, and are therefore not used except where their necessity is indicated by the state of the bodily habit; cupping, however, is not unfrequently resorted to." In the "Description of the Retreat," the preparation of which was suggested by the composition of the above letter, the author writes much to the same effect. In both he dwells upon the success attending the feeding of maniacal patients "freely with meat, or cheese and bread, and good porter."* In the "Review of the Early History of the Retreat," the same writer observes, "From the history of patients brought under care at an early period, it was evident that the reducing system had been extensively used; low diet as well as active depletory means had generally been resorted to. To the abandonment of this system must, we believe, be attributed, in no inconsiderable degree, the mental recovery of many patients, as well as the preservation of life and the increase of its comfort" (p. 34).

Let us see, now, how consistently Dr. Rush carried out his theories in regard to the seat of insanity in the cerebral vessels, in his treatment of those labouring under the disease. Among the remedies for hypochondriasis or tristimania, he observes in the work just referred to, that should the disease continue after its causes have been removed, " recourse should be had to bloodletting if the pulse be tense or full, or depressed without either fulness

---

* " Description of the Retreat," by S. Tuke, 1813, p. 118. The author, in preparing the work for a new edition, wrote, "This plan of treatment (depletion) appears to be founded on the supposition that irritation or violence proceeds universally from a plethoric habit, whereas I think experience clearly contradicts this opinion."

or tension. I have prescribed this remedy with success, and thereby in several instances suddenly prepared the way for its being cured in a few days by other medicines." Of this he gives a remarkable illustration. He also prescribes in some, if not most cases, blisters, issues, salivation, emetics, purges, and a reduced diet. Then, "after reducing the action of the blood vessels to a par of debility with the nervous system, or to borrow an allusion from a mechanical art, after *plumbing* those two systems," he allows stimulating aliment, drinks, and medicines. He observes that warm tea and coffee made weak are generally grateful to the stomach, and that Burke often relieved the low spirits from which he suffered by sipping a tea-cup full of hot water. He extols opium as "that noble medicine which has been happily called the medicine of the mind," and holds that it has many advantages over ardent spirits as a cordial. He approves of wine in the condition just described as resulting from reduction by depletion, and says he had once known the disease cured by the liberal use of Madeira. Under the head of Amenomania, "which is a higher grade of hypochondriasis, and is characterised by exaltation instead of despondency," Dr. Rush advises nearly the same remedies as in tristimania, and "particularly bleeding, purging, emetics, and low diet, in an excited state of the blood vessels, and after they are reduced, stimulating diet, drinks and medicines, and a change of company, pursuits, and climate."

As might be expected, Rush prescribes blood-letting as the first remedy in mania, which he observes is indicated by all the facts in favour of mania "being an arterial disease of great morbid excitement or inflammation in the brain," and " by the importance and delicate structure of this organ which forbid its bearing violent morbid action for a length of time without undergoing permanent obstruction or disorganization." He also held that venesection was necessary in consequence of there being " no outlet from the brain to receive the usual results of disease or inflammation, particularly the discharge of serum from the blood-vessels." He appealed to the accidental cures of insanity following the loss of large quantities of blood, and it must be granted that this has occasionally occurred in the case of patients who have cut their throats. He likewise pointed to the morbid condition of the blood in mania, and says he never saw a single instance in which it presented a natural appearance. But the final truth of his darling remedy was " the extraordinary success which has attended its artificial use in the United States, and particularly in the Pennsylvania Hospital."

It is interesting, now-a-days, to recall his special directions for bleeding in mania. It must be copious. From 20 to 40 ounces of blood are to be taken at once, unless syncope occurs before it is drawn. He asserts that " the effects of this early and copious bleeding are wonderful " in calming mad people. It often obviates the necessity of using any other remedy, and sometimes cures in a few hours. Dr. Rush even adds that bleeding must be continued, not only when there are the obvious indications referred to, but " in the absence of them all, provided great wakefulness, redness in the eyes, a ferocious countenance, and noisy and refractory behaviour continue, all of which indicate," he adds, "a highly morbid state of the brain." It is noteworthy that our author insisted that the quantity of blood drawn in mania should be greater than in any other organic disease. He selects from his successful cases of profuse bleeding a man of 68 from whom he abstracted between December 20th, 1806, and February 14th, 1807, nearly 200 ounces of blood; and another man who lost in 47 bleedings about 470 ounces. In a third case of recent madness with cool skin and natural pulse, but with insomnia and eyes suffused with blood, Rush bled him copiously, after which the pulse became frequent and tense. He repeated the bleeding, and gave him several purges, with the result that the patient recovered in a few days. Well may he say that the attention of the pupils of the hospital was attracted to this case in a more than ordinary degree. Even in "Derangement in the Memory," indicated by its weakness or loss, he recommends "depleting remedies, if plethora attend, and the pulse be tense or oppressed." After the reduction of the system by bleeding, purges, and low diet, the remedies prescribed were blisters, issues in the arms, errhines, the cold bath, and exercise.

I now proceed to note the practice of Rush's contemporaries and successors in respect to depletion in insanity. Although I shall refer especially to alienists in America, I shall cite some leading authorities in Europe. In this inquiry I am under great obligations to Dr. Pliny Earle, who some years ago carefully collected the opinions of many mental physicians in reference to bleeding. I have freely availed myself of his valuable production, "Examination of the Practice of Blood-letting in Mental Disorders," written in 1854. In it the author says, "Few truths in pathology are better established than that active sthenic inflammation is of very rare occurrence in those forms of disease ordinarily included under the general term of insanity." He also observes "What might then have been the success of a practical adherence to the therapeutics of Dr. Rush cannot now be determined except by inference. The intelligent, judicious, and successful superintendent, Dr. Kirkbride, in the early part of his career, took occasion to caution physicians against pursuing a course of treatment proper for the inflammation of the brain, and there is good authority for the supposition that in cases of mania, unaccompanied by epilepsy, paralysis, or apoplexy, or by some adventitious or accidental disease, he has never during his management of that institution* practised venesection" (p. 98).

Take the treatment recommended by an English phy-

---

* The Pennsylvania Hospital for the Insane, Philadelphia.

sician of some repute in his day who wrote in 1792, Dr. Pargeter.* He observes that as the condition present in acute mania manifestly depends on an undue and increased excitement, the object should be "to derive blood from the brain." He prescribes "abstinence to a very considerable degree," and says that "maniacs can abstain from food with wonderful perseverance." "Bleeding, copious and repeated, in the jugular vein, wonderfully mitigates," we are assured, "morbid heat, proves highly anti-spasmodic, lessens the tone of the *fibræ motrices*, and tends to prevent any topical determination. Especially useful is this mode of depletion when mania is complicated with epilepsy or hysteria."

Mr. Crowther, the visiting surgeon to Bethlem Hospital, writing in 1811† states: "I have bled 150 patients at one time, and have never found it requisite to adopt any other means of security against hemorrhage than that of sending back the patient to his accustomed confinement." Not a single instance can he adduce of deleterious consequence from the occurrence of a fresh bleeding. "The most violent I have been obliged to place on the floor and then secure them by assistants, and place myself in a like position in order to perform the operation without danger."

Haslam, also representing the practice of Bethlem at that period, says: "Where the patient is strong and of a plethoric habit, and where the disorder has not been of any long continuance, bleeding has been found of considerable advantage, and as far as I have yet observed is the most beneficial remedy that has been employed. The melancholic cases have been equally relieved with the maniacal by this mode of treatment."‡

The same authority, in his evidence before the House of Commons in 1815, said "the period of physicking con-

---

* "Observations on Maniacal Disorders."
† "Practical Remarks on Insanity."
‡ "Observations on Madness and Melancholy," 1809, p. 313.

tinued from the middle of May, regulated by the season, to the latter end of September; two bleedings, according to discretion; half-a-dozen emetics, if there should be no impediment to their exhibition; and the remainder of the time until Michaelmas, a cathartic once a week." ("Minutes of Evidence," p. 63.)

The physician to Bethlem Hospital at the same date, Dr. F. Monro, has been often quoted to show the practice of that day, but he may be cited once more for our present purpose.

"Patients," he said, "are ordered to be bled about the latter end of May or beginning of June according to the weather. After they have been bled they take vomits once a week, for a certain number of weeks; after that we purge them." ("Minutes of Evidence taken before the Select Committee of the House of Commons," 1815, p. 95.)

On these opinions Dr. Earle forcibly observes: "The practice formerly pursued at Bethlem, as revealed by Crowther, Haslam, and Monro, is so palpably and so utterly absurd as to require no comment. It is difficult to conceive how it were possible, at so recent a period as 1815, that men of professional eminence in London should still uphold so gross an error. It is a remarkable example of the adherence to traditional custom, regardless of the question of their propriety." (*Op. cit.*)

Esquirol gives a graphic description of blood-letting as the unhappy result of Harvey's great discovery. He writes in 1816:—

"On the discovery of the circulation of the blood, it was believed that we had discovered the cause of every disorder, and a remedy for all ills. Blood was shed abundantly. The blood of the insane was the more freely shed, as by bleeding them to faintness it was believed that they were cured. This treatment was extended to all the insane. In every asylum there was established what was called the treatment of the insane, on this principle: that the blood being too abundant, or too much heated, ought to be evacuated and cooled. Besides, in the asylums of France, where some attention was paid to the insane, in spring and autumn they bled them once or twice, and bathed them in cold water; or cast them, bound hand and foot, into a river or reservoir. If a few victims of such gross mismanagement escaped, they cried out, 'a miracle!' Such was the prejudice not long since, even in Paris, in favour of bleeding, that we were accustomed to receive pregnant women, who had been bled by way of precaution before being sent to a house where bleeding was prescribed. Excess in this respect has sometimes been so great that I have had in charge an insane man who had been bled thirteen times in forty-eight hours. Pinel set himself against this abuse, and cites examples which ought to be present to the mind of all physicians. I can add, that I have many times seen insanity increase after abundant hemorrhages of various kinds, and after one, two, and even three bleedings. I have seen a state of sadness pass into mania and fury, immediately after bleeding; and dementia to replace, reciprocally, the condition of mania. I do not believe it necessary to prescribe blood-letting in the treatment of insanity. It is indispensable in plethoric subjects, when the head is strongly congested, and hemorrhages, or habitual sanguine discharges, have been suppressed. At the commencement of insanity, if there is plethora, if the blood rushes violently to the head, if some habitual hemorrhage is suppressed, we bleed largely, once, twice, or thrice, apply leeches over the jugular veins and temporal arteries, and cupping glasses to the back of the head. At a later period sanguine discharges are local, and employed as revulsives or as supplementary to suppress evacuations, etc." *

Sir Alexander Morison, while protesting against the indiscriminate employing of blood-letting, observes that it is "absolutely necessary in some cases." †

* "Des Maladies Mentales," Tome i., pp. 151-3.
† "Outlines of Lectures on Mental Diseases," 1826, p. 86.

Writing as late as 1828, at least 30 years after venesection had been practically discarded at the York Retreat, Dr. George Man Burrows stated that "copious abstractions of blood are almost universally adopted in cases of insanity attended with symptoms of violence, and sometimes when the patient is tranquil. The practice has received the sanction of ancient authority, and is at present very universal." Dr. Burrows then asserts that, unless in very exceptional cases, it is a practice fraught generally with mischief, and he makes a confession honourable to himself and of great service to our inquiry: "Following example rather than experience, I tried depletion by blood-letting for several years; but discovering my error, I became more cautious, and I believe I have scarcely ordered venesection in six cases of simple mania or melancholia in as many years. My conclusion is that since I changed my practice more have recovered, and certainly the cases have been less tedious and intractable."* At the same time he rejected the sweeping condemnation of the lancet and cupping in mental derangement, and thought that both *might* be required. Indeed, he believed that leeching "could seldom be dispensed with in any recent case." Nay, speaking of his own experience, he says: "In *every* case of recent insanity which I have seen, local abstraction of blood has been indicated. (*Op. cit.*, pp. 587-589.)

Prichard observes: "I am very far from approving or wishing to recommend such abstractions of blood as those which appear to have been practised by Dr. Rush; but I have been convinced by evidence of numerous facts that bleeding, both local and general, is under due limitations serviceable in cases of insanity." †

All the references which follow are from the writings of Rush's own countrymen. Writing in 1841 Dr. Luther Bell, the distinguished Superintendent of the McLean

* "Commentaries on the Causes, &c., of Insanity," 1828, p. 583.
† "On Insanity and other Disorders affecting the Mind," 1837.

Asylum, Boston, asserted that no individual at the head of an insane institution would now think of combating any form of insanity with the depletory and reducing means once regarded as indispensable. ("Annual Report.")

The practice of another American alienist, Dr. Todd, formerly of the Hartford Retreat, Connecticut, is recorded by Dr. Brigham in the following words:—

"He early discountenanced depletion, particularly bleeding, in insanity, and insisted upon the necessity of generous diet, and recommended a frequent resort to tonics and narcotics in the medical treatment of the insane. This course of treatment, though it had been recommended by the best writers on insanity in Europe, had not, to much extent, been resorted to in this country previous to the time of Dr. Todd, and it was contrary to that recommended by Dr. Rush, that it required considerable boldness and much address and management to introduce it and make it popular in this country, and this Dr. Todd accomplished."*

Dr. Brigham himself (formerly Superintendent of the Utica Asylum) observes, in his Annual Report, 1847: "The treatment of insanity by bleeding, though strongly recommended by Dr. Rush and some others, we believe to be generally improper and frequently very injurious. In some it produces a fatal result, and, we are confident, not unfrequently renders cases incurable."

An American, Dr. Samuel White,† of the Hudson Asylum, New York, spoke as follows in an address on "Insanity" before the New York Medical Society:—

"With great deference I have ventured to reverse the order of Dr. Rush, who says 'the cause of madness is seated primarily in the blood-vessels of the brain.'

"This theory, I apprehend, has too often in incipient

---

* "American Journal of Insanity," Vol. vi.
† Vice-President of the American Association of Medical Superintendents of American Institutions for the Insane, 1844 to 1846.

insanity led young practitioners into the fatal error of treating by bold depletion *irritation* for *inflammation,* than which, as a general rule, nothing is more prejudicial to the radical cure of the patient."

I will next take the testimony of Dr. Ray, who, in the third Annual Report of the Maine Hospital for the Insane (1843), observes :—

"I trust that I may be allowed without offence to call the attention of my professional brethren, who are so partial to the depletory treatment advocated by Rush and certain English writers, to the important fact that in no hospital for the insane in New England—and the same may be the case in many other institutions in our country —is this treatment now used. I cannot but think that if this fact were generally known, it would lead to a thorough revision of the ground of their treatment, or, at least, would render them more cautious how they practise those enormous abstractions of blood which so often lay the foundation of hopeless fatuity or a tedious convalescence."

Lastly, I would quote Dr. Curwen, for many years the Superintendent of the State Lunatic Hospital at Harrisburg, and now occupying the same position at Warren, who, in his Report of 1852, observes :—

"I feel I am discharging a part of my duty towards the insane in calling attention to an error which is very extensively prevalent, and consists in the almost invariable resort to blood-letting in all cases of insanity."

This is a very remarkable statement, and would, perhaps, rather tend to show that the change in the treatment of the insane by blood-letting, which marked the practice of some physicians, was the result of a change of opinion rather than of the type of disease from sthenic to asthenic, for here we have the majority of those engaged in that branch of medicine following, in 1852, the practice of Dr. Rush in as full a belief in its efficacy as was entertained by Dr. Rush in 1812. And it may be observed that neither Dr. Ray nor others who protested against the sanguinary treatment pursued by their fellow-practitioners, contended that such change in the form of disease had taken place. They appear to have thought that depletion was, and always had been bad, not only when they wrote, but when Rush practised it.

I will now refer to the position taken by Rush in regard to Moral Insanity.

"I once knew a man," says he, "who discovered no one mark of reason, who possessed the moral sense or faculty in so high a degree that he spent his whole life in acts of benevolence. He was not only inoffensive (which is not always the case with idiots) but he was kind and affectionate to everybody. He had no ideas of time but what were suggested to him by the returns of the stated periods for public worship, in which he appeared to take great delight. He spent several hours of every day in devotion, in which he was so careful to be private that he was once found in the most improbable place in the world for that purpose, viz., in an oven."

This case is from an essay written by Dr. Rush exactly a century ago, entitled, "An Inquiry into the Influence of Physical Causes upon the Moral Faculty;" and I confess as I read it and a chapter of his work on the Mind in which he speaks of "moral derangement and of innate preternatural moral depravity, the result of original defective organization," I am driven to ask myself whether we do not even now require to be taught the truths which its author enforces with so much power and originality. Is it that in the study of the minute shades of difference between the various forms of insanity and the multitudinous questions which have presented themselves for solution since his day, we are unable to recognise the broad character and bold outline so lucidly sketched by the hand of this great master? Have we become unable to see the forest for the trees? What should we think of the physician who, when shown a case of congenital alopecia, should, after examining it with minute accuracy, discover one or two stray hairs on the scalp and thereupon declare that it was no case of alopecia at all? Surely this would be to miss the essential feature of the case in noting a mere trifling exception. Let us discover, then, if we can, whatever intellectual flaw there may be—however slight—in every instance in which the emotions are primarily and essentially the seat of disorder, but let us not, in doing so, lose the lesson which the case really teaches.

It has often seemed to me strangely inconsistent that, while we demur, on psychological grounds, to the separation of mental functions which is implied in the doctrine of the lesion of the emotions without disorder of the intellect, we accept as nothing at all extraordinary the converse condition of healthy affections with morbid derangement of some intellectual function. What stickler for the unity and solidarity of the mind hesitates to believe that our memory may fail, and yet our love remain the same; that our power of attention may become extremely weak, while there is no increased tendency to depravity? I have known a man lose his knowledge of several languages in a railway accident, but he kept the Ten Commandments just as well after as before this intellectual loss. This will not be disputed. Then why should it be thought metaphysically heterodox to believe that an injury to the head can alter the moral character without a perceptible weakening of the memory, the attention, or any other intellectual faculty?

Some years ago I was struck with the remarkable case of Servin, described in Sully's "Memoirs of Henry IV. of France," as possibly one of moral insanity or moral imbecility, and was not aware that Rush had made use of this extraordinary character to illustrate his theme. The chief question which arises in this class of cases is not —Was there any intellectual aberration, but rather—Are we justified in regarding such moral monsters as anything more than horribly vicious and responsible for their acts? Rush regarded Servin as a case of "universal moral derangement." Sully says: "Just before my departure for Calais, old Servin came and presented his son to me, and begged I would use my endeavours to make him a man of some worth and honesty; but he confessed it was what he dared not hope, not through any want of understanding or capacity in the young man, but from his natural inclination to all kinds of vice. The old man was in the right; for what he told me having excited my curiosity to gain a thorough knowledge of young Servin, I found him to be at once both a wonder and a monster; for I can give no other idea of that assemblage of the most excellent and most pernicious qualities. Let the reader represent to himself a man of a genius so lively, and an understanding so extensive, as rendered him acquainted with almost everything that could be known; of so vast and ready a comprehension, that he immediately made himself master of whatever he attempted, and of so prodigious a memory that he never forgot what he had once learned. He possessed all parts of philosophy and the mathematics, particularly fortification and drawing; even in theology he was so well skilled that he was an excellent preacher whenever he had a mind to exert that talent, and an able disputant for and against the reformed religion indifferently. He not only understood Greek, Hebrew, and all the languages which we call learned, but also all the different jargons or modern dialects; he accented and pronounced them so naturally, and so perfectly imitated the gestures and manners both of the several nations of Europe and the particular provinces of France, that he might have been taken for a native of all or any of these countries; and this quality he applied to counterfeit all sorts of persons, wherein he succeeded wonderfully; he was moreover the best comedian and greatest droll that perhaps ever appeared; he had a genius for poetry, and had written many verses; he played upon almost all instruments, was a perfect master of music, and sang most agreeably and justly; he likewise could say mass, for he was of a disposition to do as well as to know all things; his body was perfectly well suited to his mind, he was light, nimble, dexterous, and fit for all exercises; he could ride well, and in dancing, wrestling, and leaping, he was admired; there are no games of recreation that he did not know; and he was skilled in almost all mechanic arts.

"But now for the reverse of the medal. Here it appeared that he was treacherous, cruel, cowardly, deceitful; a liar, a cheat, a drunkard, and glutton; a sharper in play, immersed in every species of vice, a blasphemer, an atheist; in a word, in him might be found all the vices contrary to nature, honour, religion, and society; the truth of which he himself evinced with his latest breath, for he died in the flower of his age, in a low house, perfectly corrupted by his debaucheries, and expired with a glass in his hand, cursing and denying God." *

It may well, I say, be a question in this instance whether controllable vice, or uncontrollable impulses, would the more properly characterize the case of this young man. I think we may reasonably conclude that there was some constitutional peculiarity, while I should hesitate to pronounce him wholly irresponsible. We may, however, agree with Rush, that "such persons are, in a pre-eminent degree, objects of compassion, and that it is the business of medicine to aid both religion and law in preventing and curing their moral alienation of mind." (*Op. cit.*, p. 358.)

Rush could hardly be expected to write with complete perspicuity in all he says, when he treats on this subject. He seems to contradict his own principles to some extent when he discusses moral responsibility. His use of the words "vice" and "wickedness" is consequently loose and undefined. The wonder, however, is, not that he allowed himself to be betrayed into inconsistencies of expression, but that he saw so far in advance of the age in which he lived, and addressed himself so forcibly to it.

* "The Memoirs of the Duke of Sully, Prime Minister to Henry the Great," Vol. iii., p. 35.

Rush has been styled the American Sydenham. I should rather call him the American Fothergill. He resembles this physician (whom he personally knew and admired) in the independence of his medical practice; in acuteness of observation; in an enthusiastic love of the art of healing; in his incessant labour; in popularity as the leading physician of the day in a great city; but above all, in uniting with the functions of a physician the philanthropy which manifested itself in innumerable practical suggestions for the benefit of his kind, and in the daily exemplification of Terence's immortal axiom. Like Fothergill he did not practise to live, but he lived to practise. Like Fothergill, also, he might have said in relation to diseases what the Roman said in relation to battles: "In the midst of them I have always found time to contemplate the stars, the tracts of heaven, and the realms above." Such men serve to raise the character of the profession above all mercenary considerations; they recognise that their calling is divine; and they afresh illustrate the truth of Cicero's proposition, "*Homines ad deos nullâ re propius accedunt quam salutem hominibus dando.*"

I must now draw this somewhat desultory notice of a remarkable physician and excellent man to a close. I am conscious that the citations from his writings fail to do him justice. I had almost said they do him injustice, because it requires a study of the whole of his writings to convey a faithful picture of his opinions. It is true that, if we take isolated passages from his work on Insanity, he appears to disadvantage; but a perusal of the whole, while it shows that he was not free from some of the strange notions then prevalent in regard to the treatment of the insane, leaves the conviction upon the mind of the reader that he was an original observer, a humanely intentioned, and in many instances a successful, physician of the insane. It cannot, I think, be denied that those who contemporaneously with him were endeavouring in Paris and in York to ameliorate the condition of the insane, saw further than he did as to what could and ought to be done; and that it would be easy to illustrate this by a series of parallel passages from the works which describe the moral and medical treatment pursued at the Bicêtre and the York Retreat. Still we cannot withhold our admiration from the lancet-loving physician of Philadelphia, who, amidst his incessant engagements and varied practice, studied mental disorders profoundly, advanced in this track far beyond his teacher, Cullen, and found time to compose a monograph on derangement of the mind, which, had he written nothing else, would have given him an enduring name in the republic of medical letters.

The reflections on the death of Prichard made by a kindred soul, Dr. Symonds of Bristol, are so applicable to that of Rush, that I offer no apology for citing them as an appropriate conclusion to this brief sketch:—

"Though he had not ceased from his labours, nay, the sickle was in his hand when it dropped, few could so well have said, though he would have been the last to say it, 'I have not lived in vain.' If one could venture in imagination to follow the musings of that departing spirit, one might conceive the satisfaction with which he looked back upon his well-spent life. . . . Youth had found him assiduous in acquiring truth and knowledge; manhood and advancing age had witnessed untiring exertions in a profession which, whatever it may produce to the practitioner, is, if grounded on adequate knowledge, an employment pre-eminently useful to his fellow-creatures. And the intervals in those avocations, instead of having been set apart, as they might innocently have been, for recreation and amusement, had been filled up with labour which, had he done nothing else, would have enabled him to bequeath honour to his family, as the inheritors of his renown, and lasting benefits to mankind of the highest order, for I know not what gifts can surpass those of truth and wisdom. . . . And one fancies that, with such remembrances, he might well say, *Nunc dimittis*. . . . But

I doubt not that the deeds of his life, which to us look large and brilliant, before *his* failing sight shrank small and dim, and that his soul, which no earthly vision could content, much less the contemplation of his own doings, turned towards that Parent Source from which all his light had been drawn, and longed to be absorbed into its divine and immortal essence. Though, however, he would have depreciated rather than magnified himself, we who look at him from without, and estimate him by the standards that enable men not only to recognise moral excellence, but to mete out the degrees of their approval, cannot refrain from declaring that no spirit could pass more blameless and unstained from its mortal trial, none more fitted for the communion of the great and good, none more ready to appear

> 'Before the Judge who henceforth bade him rest,
> And drink his fill of pure immortal streams.' " *

---

* Meeting of the Bath and Bristol Branch of the Provincial Medical and Surgical Association, March, 1849.

CHAPTER II.

PROVISION FOR THE INSANE IN THE UNITED STATES FROM 1752 TO 1876.

IT may not be uninteresting to the readers of this book to have a slight historical sketch of what may be called the past asylum movement in the States. I am not aware that this has been given before in any journal or work published in Great Britain. The peculiar difficulties of a new country, peopled by different races, and the constantly disturbing influence of immigration, ought to be borne in mind in this narrative. These difficulties are too frequently overlooked. In a letter I received from Miss Dix in 1874, she wrote—"We have an amazing burden in all our charitable institutions of every class of disabled foreigners of all ages and in all stages of feeble or quite broken-down conditions of health."

As in England, so, no doubt, in America, frightful abuses have existed—more than that, much remains to be done.* The insane have been subjected to the same barbarous neglect and treatment as with us. Puritanism, in the first instance, was only too likely to treat some forms of madness as instances of witchcraft, and their subjects would be punished or put to death accordingly. Other cases would be simply referred to the cruel action of Satan upon the mind, and proper medical treatment would be the last thing thought of. A good illustration of the belief in such diabolical influence in mental depression occurs in Cotton Mather's "Life of William Thompson." "Satan," he says, "who had been often in an extraordinary manner irritated by the evangelic labours

* That is, at the close of the period to which this chapter refers, namely 1876—the date at which this sketch was written.

of this holy man, obtained the liberty to sift him; and hence, after this worthy man had served the Lord Jesus Christ in the church of our New English Braintree, he fell into that *balneum diaboli*—a black melancholy, which for divers years almost wholly disabled him for the exercise of his ministry." .... "New England, a country where splenetic maladies are prevailing and pernicious—perhaps, above any other—hath afforded numberless instances of even pious people, who have contracted those *melancholy indispositions*, which have unhinged them from all service or comfort. Yea, not a few persons have been hurried thereby to lay violent hands upon themselves at the last. These are among the unsearchable judgments of God." When they were really regarded as madmen, the care and treatment of the insane were probably neither better nor worse than in the mother countries from which the early settlers came. Humanity and Science, however, attacked and at last broke through the strongholds of superstition, ignorance, and prejudice; and benevolent men (and women, too) exerted themselves to mitigate the unhappy condition of those who were confined as lunatics, and to provide for them suitable accommodation and kinder treatment.

It appears that it was from the Province of Pennsylvania that the first humane impulse proceeded. It was fitting that the State founded by the humane and enlightened Penn, should take the lead in this work of mercy. From Philadelphia there went a petition to the House of Representatives, in which it is stated that, with the increase of the population, the number of the insane has greatly increased; "that some of them going at large are a terror to their neighbours, who are daily apprehensive of the violence they may commit; and others are continually wasting their substance, to the great injury of themselves and families—ill-disposed persons wickedly taking advantage of their unhappy condition, and drawing them into unreasonable bargains—that few of them are

so sensible of their condition as to submit voluntarily to
the treatment their respective cases require, and therefore
continue in the same deplorable state during their lives."
The House is requested to aid in founding a small provincial hospital for these and other persons labouring
under disease; for it seems that in the first instance it
was not designed exclusively for the insane. This, the
petitioners affirm, will be "a good work acceptable to
God and to all the good people they represent." This
was in 1751. The consequence was that the Legislature
passed the necessary Act; a sum of money was voted,
subject to an equal amount being raised by private means,
and the new hospital was opened at Philadelphia in the
following year. Although anticipating the course of
events, it should be added that in 1841 the insane patients
were transferred to the new "Pennsylvania Hospital for
the Insane," near Philadelphia, of which Dr. Kirkbride
was appointed Superintendent.

It is pointed out by Dr. Ray* (to whom I am indebted
for these and other particulars) that this Pennsylvania
Hospital has an additional claim on our gratitude, inasmuch as it was here that Dr. Rush obtained the materials
for his work on "Diseases of the Mind," published in
1812.

Twenty-one years later (1773) Virginia established at
Williamsburg an asylum, or, as the Americans very properly call such institutions, a "hospital," which provided
for the insane only. "Many years elapsed," Dr. Ray
states, "before this worthy example was followed; nor
was the great want supplied by associations like that
which founded the Pennsylvania Hospital, nor by individuals, as in the private asylums of England. The latter
class of enterprises was almost unknown in this country
until the beginning of the present century; for they
required a knowledge of insanity not easily obtained by our

---

* Address on the occasion of laying the foundation stone of the Danville
Hospital for the Insane (Penn.), in 1869.

physicians, an outlay of capital which few of them possessed, and a rate of prices greatly beyond the means of our people. In process of time they made their appearance, few and far between, but their benefits were confined to the affluent classes."

In 1817 the Friends established an asylum at Frankford, near Philadelphia, with the object of carrying out the system of treatment pursued at the York Retreat. In the following year the McLean Asylum, at Somerville, Mass., was opened. Dr. Wyman, the first physician, appears to have devoted himself warmly to the interests of the patients. He opposed the indiscriminate use of bleeding, purging, and low diet. The Bloomingdale Asylum (the lunatic department of the New York City Hospital) was opened in 1821. At one period Dr. Pliny Earle, now the physician-in-chief of the Northampton Hospital for the Insane, Mass., superintended this asylum. The Hartford Retreat, Connecticut, opened in 1824, was first superintended by Dr. Todd, who, like Dr. Wyman, objected to depletion, and employed tonics, sedatives, and a generous diet. Dr. Brigham was superintendent from 1839 to 1843. He was succeeded by Dr. Butler. In 1821 an appropriation was made by the South Carolina Legislature towards an asylum at Columbia, which was completed in 1827.

It is worthy of remark that in no State in America, Virginia excepted, did the Legislature undertake wholly to provide for the insane until 1832. In that year Massachusetts erected an asylum at Worcester (the first Superintendent being Dr. Samuel B. Woodward), and from this time it was regarded as a duty for the States not merely—as in the instance of Pennsylvania—to aid the efforts of others, but to provide the *whole* amount required for the erection of hospitals for the insane. The just dictum of Horace Mann, that insane paupers are the wards of the State, is regarded by Ray as having "taught the people, now and for ever, the exact nature of their relations to this class of their fellow creatures. . . . . The example of Massachusetts, executed as well as conceived in a most generous manner, was followed by other States, one after another."

There appear to have been several causes for the movement which now happily commenced. One was that the experience already gained in the hospitals, where the insane were judiciously and kindly treated, had proved that they might and ought to be removed from jails and poorhouses, and placed under proper care and treatment. For this State aid was indispensable. Another cause was that with the Hour, there came fortunately the Man, or rather the Woman; for, at this juncture, a lady whose attention was directed to the condition of the insane, resolved to devote herself to their service; and from that day to the present time, the insane, if neglected or ill-treated, have had in Miss Dix a powerful and untiring advocate. From his personal knowledge of this philanthropic lady's character, the writer can well believe the statement made by the physician already cited, that " no place was so distant, no circumstances so repulsive, no lack of welcome so obvious, as to deter her from the thorough performance of her mission. Neither the storms of winter, nor the heats of summer, could diminish the ardour of her zeal, and no kind of discouragement could prevent her from gauging exactly the dimensions of this particular form of human misery;" and he adds, that "favoured by that exquisite tact and happy address peculiar to her sex, she overcame obstacles that would have defied the ruder efforts of the other sex, and thus brought to light a mass of suffering that seemed more like an extravagant fiction than real unexaggerated truth." Miss Dix appealed to the Legislatures of the various States to pass enactments calculated to remedy this state of things, and terminate " practices that would shock even a barbarous people." Her efforts were generally rewarded by the desired action being taken to provide hospitals for the insane. But, even in the "go-ahead" land across the Atlantic, the interval which elapsed between the decision to act and the act itself, was provokingly long, and it required the watchful eye of the promoter of these benevolent measures, aided by a few medical men, to ensure their being really carried into execution.

It was in 1848 that Miss Dix presented a Memorial to the Senate and House of Representatives of the United States in regard to the necessity for the relief and support of the indigent curable and incurable insane. It contains harrowing descriptions of their condition in private houses, almshouses, and jails. Some were in cages, filthy cells, and very many in chains and irons, "reduced to the most abject moral, physical, and mental prostration."

In the State of New York Miss Dix found that nearly every poor-house had its crazy-house, crazy-cells, crazy-dungeons, or crazy-hall. "At A——, in the cell first opened, was a madman. The fierce command of his keeper brought him to the door, a hideous object; matted locks, an unshorn beard, a wild, wan countenance, disfigured by vilest uncleanness; in a state of nudity, save the irritating incrustations derived from that dungeon, reeking with loathsome filth. There, *without light*, without pure air, without warmth, without cleansing, absolutely destitute of everything securing comfort or decency, was a human being—forlorn, abject, and disgusting, it is true, but not the less a human being—nay, more, an immortal being, though the mind was fallen in ruins, and the soul was clothed in darkness. . . . Is your refinement shocked by these statements? There is but one remedy; the multiplication of well-organised hospitals, and to this end, creating increased means for their support."

The following are among the noble words of this appeal to the Legislature:—

"I advocate the cause of the much-suffering insane throughout the entire length and breadth of my country.

I ask relief for the east and for the west, for the north and the south; and for all I claim equal and proportionate benefits.

"I ask of the Senate and House of Representatives of the United States, with respectful but earnest importunity, assistance to the several States of the Union in providing *appropriate care and support for the curable and incurable indigent insane.*

"I ask of the representatives of a whole nation, benefits for all their constituents. Much has been done, but much more remains to be accomplished, for the relief of the sufferings and oppressions of that large class of the distressed for whom I plead, and upon whose condition I am solicitous to fix your attention.

"I ask for the people that which is already the property of the people; but possessions so holden, that it is through your action alone they can be applied as is now urged.

"The whole public good must be sought and advanced through those channels which most certainly contribute to the moral elevation and true dignity of a great people.

"I will not presume to dictate to those in whose humane dispositions I have faith, and whose wisdom I cannot question.

"I confide to you the cause and the claims of the destitute and of the desolate, without fear or distrust. I ask for the thirty States of the Union 5,000,000 acres of land, of the many hundreds of millions of public lands, appropriated in such a manner as shall assure the greatest benefits to all who are in circumstances of extreme necessity, and who, through the Providence of God, are *wards of the nation*, claimants on the sympathy and care of the public, through the miseries and disqualifications brought upon them by the sorest afflictions with which humanity can be visited."

To resume. While the idea of founding a State asylum in Pennsylvania was projected in 1838, and an Act was obtained, the project fell through; and it was not till 1845 that a successful attempt was made to obtain another Act, which, strange to say, was itself not carried into effect for years, and the asylum (at Harrisburg) was not opened before 1851. An insane hospital, the funds necessary for which were partly provided by private individuals and partly by the State, was also opened in 1861, for Western Pennsylvania, and called the Dixmount Hospital.

In 1836, or a little later, "the attention of certain philanthropic and enlightened citizens of New Hampshire began to be turned towards some better provision, or rather towards some provision, for its insane. The success of the State Lunatic Hospital at Worcester, in the adjoining State, was rapidly being recognised, and the enquiries, set on foot by Dr. Bell and his associates, revealed an amount of suffering before unsuspected. Among those who devoted themselves to this thankless and unpopular effort to induce the community to wake from its guilty lethargy, deserve to be enumerated the names of General Peaslee, President Pierce, S. E. Cones (now of Washington), the late Charles J. Fox, and a few others. Time after time the Legislature refused the necessary sanction for an asylum. Eventually, however, these efforts proved successful, and resulted in the establishment, by private subscriptions and State aid, of that excellent institution, the New Hampshire Asylum for the Insane."*

Dr. Bell reported the number and condition of the insane in this State, and the means of providing for them; and his report was not only ordered to be published for distribution by the Legislature, but was reprinted in the Journals of both Houses as worthy of perpetuation in the governmental history of New Hampshire. (Shortly after, he was appointed physician-superintendent of the McLean Asylum, Boston, Mass.) The New Hampshire Asylum

---

* "American Journal of Insanity."

was opened in 1842, Dr. Chandler being appointed Superintendent. In 1845, Bell visited Europe, and found, as he said he had expected to find, that much progress had been made in the construction of asylums in Great Britain. Of this he availed himself, and prepared plans for the erection of the Butler Hospital for the Insane, at Providence, Rhode Island; of which institution Dr. Ray was appointed the first superintendent. He had previously superintended the Maine Asylum, opened in 1840. The New York State Lunatic Asylum, Utica, was opened in 1843. The well-known Dr. Brigham was superintendent from its opening until his death in 1849. Dr. Gray now (1876) fills that office. In 1844 Miss Dix induced the Legislature of New Jersey to take up the question of provision for the insane in that State, and to appoint a committee to select a suitable site for a building. Dr. Buttolph was appointed medical superintendent.

On reference to the admirable Report on the insane in Massachusetts, drawn up by Dr. Jarvis, and presented in 1855, we learn that there was at that time 1 lunatic in every 427, and 1 idiot in every 1034 of the population, or 1 of either class in 302. There were 2632 lunatics, 1087 idiots; of the former, 1284 were at their homes or in town or city poorhouses; 1141 in hospitals; 207 in receptacles for the insane, in houses of correction, jails, and State almshouses. Of the latter, 670 were supported by friends, and 417 by the public treasury. The pauper class of lunatics, it is stated, furnished in ratio of its numbers sixty-four times as many cases of insanity as the independent class.

In 1856 Dr. Bell said in his Report, "The number of Hospitals for the Insane in the United States has increased during the last 19 years from 6 to between 40 and 50, and the accommodation for patients has risen from about 500 to between 10 and 11,000. Even the four larger British provinces adjoining us have caught the influence of our zeal, and each of them has, during that period, provided itself with a large and well-furnished institution, essentially upon our models."

(Dr. Bell wrote very strongly, I may remark in parenthesis against the association of the sexes in asylums, against the frequent visits of relatives to patients, and against giving up all mechanical restraint.)

I must not, however, enter into further detail. Suffice it to say that sooner or later buildings were erected in the States, adapted for the purpose, and what is still more important, were provided with medical superintendents, devoted to their work. Among these are not a few who have distinguished themselves in this specialty, and have exerted an important influence upon medical psychology and the jurisprudence of insanity, beyond their own immediate circle. There have been features of the American asylums, I do not hesitate to say, which have been well-deserving of the attention of English physicians. The Reports of their superintendents have been and are valued by alienists in the mother country.

Dr. Woodward, so long ago as 1833, urged that, with many, intemperance was a disease requiring special care, and the American psychologists worked at this subject until an Act was passed in 1855 by the New York Legislature incorporating an association with powers to carry out this view in a definite form.* They have also taken a prominent place in the education of idiots. The late Dr. Howe, known everywhere as the enlightened friend of the idiot, was a member of a commission appointed by the Legislature, in 1846, to enquire into the condition of the idiots of Massachusetts, and to ascertain whether anything could be done for their relief. Dr. Wilbur's labours in the education of idiots are also well known. He was appointed superintendent of the New York State Institution for Idiots, in 1852.

* *Vide* "American Journal of Insanity," July, 1856. When this was written I was not aware that long before, Dr. Rush had advocated separate institutions for dipsomaniacs. (See page 12.)

It is admitted that the provision made for the insane, in at least some of the States of America, was recently and probably still is (1876) far from complete. We have referred to Massachusetts. In 1869 it was estimated that there were in this State about 2000 insane or idiotic persons unprovided for. The number of lunatics unprovided for in the State of New York some years ago, and placed in workhouses and gaols, in a deplorable condition, attracted much attention and just criticism. In 1856 there were 900 insane poor in the poorhouses and gaols of this State, 300 of whom were in cells and mechanical restraint, from one end of the year to the other. My authority is the "American Journal of Insanity."

In 1868 the Pennsylvania Medical Society memorialised the General Assembly, alleging the insufficient accommodation which existed in that State, and asserting that "a large proportion of insane persons are kept under conditions shocking to the dullest sense of propriety, or even of common humanity, suffering from cold or heat, from bad air, or indecent exposure; chained to the floor, perhaps deprived of every means of recreation or employment, and dying by that process of decay which physicians call dementia." They urged the immediate establishment of a hospital for the district, composed of the counties of Wayne, Susquehanna, Wyoming, Luzerne, Columbia, Montour, Sullivan, Bradford, Lycoming, Tioga, Clinton, Centre, Clearfield, Elk, Cameron, M'Kean, Potter, and Forest; and nine others, should the finances of Pennsylvania allow of it. The result was the erection of the Hospital for the Insane at Danville, Penn., the corner stone of which was laid in 1869, on which occasion Dr. Ray delivered the excellent address, from which we have quoted. The proper persons, we would here remark, deserving of blame for the deplorable condition of the insane, wherever it has existed or still exists in America, are not the body of alienist physicians, but the mass of the people themselves.

In 1873 were published a "Report of Public Charities in Penn.," "A Plea for the Insane in Prisons and Poorhouses in Penn.," and subsequently "Addenda," wherein evidence is given, apparently conclusive, that some at least of the insane inmates (though the proportion to the whole number may be small) were often greatly neglected and ill-treated. The condition of these, in fact, recalls that of the insane before any reforms were introduced. One old man is described as starved to death, medicine being forced into him, but food thought unnecessary; a young lady, for the last two years occupying a filthy cell, resting like a beast upon her haunches, and so permanently cramped as to be only capable of frog-like movements; a "splendid old man" in chains for 40 years, &c., &c. The Report of the Board of Public Charities gives many deplorable cases of a similar or even worse character. In one almshouse we read, "We found the female insane department in a shocking condition; so bad that it would be impossible to give a description of the place on paper. In some cells there were two or more women confined; some without any clothing, lying on the floor without mattress, carpet, or anything else, except an old Government blanket. The place had a horrible putrid odour." Of another establishment, the report is made, "Insane totally neglected, morally, physically, and medically; less attention is given to them than would be given to the lowest animals." We reproduce the above descriptions, not from any wish to throw odium upon the people of Pennsylvania for past errors, but as historical facts which we are bound to chronicle; and also as forming an instructive lesson for the future, showing, as it does, how possible it is in the midst of an enlightened community for a fearful state of things like this to remain for so long unremedied, in spite of the protests of medical men and others, and how absolutely necessary is unremitting attention to the condition of a class unable to make their own wants and sufferings known. We doubt not that much remains to be done, for at the time of which we speak (1873) it was stated that there were twice as many of the insane poor languishing in the poorhouses and prisons of Pennsylvania, as there were when Miss Dix made that appeal for their relief, in consequence of which the Harrisburg Hospital was built. Still, the Report of the Board of Public Charities for 1874, published in 1875, says—" We do not propose to detail again the sickening minutiæ of our investigations. Some of these, we thankfully believe, are buried in the dead past."

The same Report states that the number of indigent insane in the State Hospitals, established primarily for this class, "was, on 30th September, 1874, 764; the number of the same class in the poorhouses of the State and other county provision being 1,352, exclusive of 1,075 in the Philadelphia Almshouse."

The obvious remedy would seem to be the provision of a larger number of State Hospitals for the insane. Probably the obtuseness of the German element of the population has rendered Pennsylvania slower than she otherwise would have been to recognise her duties to the insane. Many of the citizens of Philadelphia (including Dr. Ray and Dr. Kirkbride) petitioned the Legislature in 1874, that "hospitals enough for the care and treatment of all the insane in Pennsylvania be prepared at the earliest possible time," and represented that "the course proposed will relieve the Commonwealth of the reproach of having insane men and women confined in almhouses, gaols, penitentiaries, or, what is worse often than either, put out of observation, neglected and inhumanly treated at their own homes, or in detached buildings near them."

I regret that in the observations of the Board of Public Charities this is not insisted upon. If, indeed, the hospitals already in existence were built for the indigent insane alone, they are right in their complaint that they are now partly occupied by those whose friends can pay for them, however moderately; or if, built for both classes, the State designed the poor to have the first claim for admission, *irrespective of curability*, the complaint is just. The law on this point is vague; for while it provides that the poor are to have precedence of the rich, it requires also that recent cases shall have precedence over those of long standing. That it admits of the construction put upon it—that a recent case, although not a pauper, shall be admitted before a chronic pauper case—seems clear from the fact that the Board of Public Charities urges upon the Legislature a more definite law on the subject. The mixture of different classes, however, in the same building would seem to be, so far, a recognised plan with the Americans; and granting this, we imagine that the question the Medical Superintendent has supposed himself bound to consider is—Which of two cases who apply for admission is the more likely to be benefited by treatment? The Medical Superintendents would seem the last persons to blame; yet, unfortunately, the tendency on the part of the Board of Public Charities appears to be to cast the odium of the state of things we have described upon them. And, further, is not some weight to be allowed to the consideration that many who are admitted on moderate terms would become paupers in a short time if not so admitted? If the Medical Superintendents are to be blamed, should not some blame be attached also to the Board of Public Charities for allowing the poorhouses to be in so bad a condition? Could not their visitation have been made more effective sooner? Ought not the Board to have done long ago what they did in 1873? Be this, however, as it may, as their proceedings have in the end been productive of good, we rejoice, although they may in some respects have erred in judgment.

In a letter I received from Dr. Ray in the summer of 1873, he states that at that time every State in the Union, excepting Delaware, and one or two of the newest States, had one or more hospitals for the insane, and they were all liberally supported in most respects. "Some of the officers think they themselves are meanly paid; and I suppose they are, in some Western and in all the Southern States. The Western—Ohio, Iowa, Illinois, Indiana—are steadily increasing their hospital capacity by building new hospitals, or adding to the old ones. In them the essential objects of such institutions, I think, are pretty well obtained, though an Englishman would probably observe some laxity in the service. In the Atlantic States, excepting New England, it is impossible to obtain good attendants, and this evil seems to be increasing every year, and the consequences are an increase of suicides, elopements, and other casualties."

A year later the same correspondent informed me that another asylum had been commenced at Warren,* in the north-west part of Pennsylvania; that in New England the hospital capacity was nearly up to the demand, and when hospitals in building were completed, no patient need be left in the poorhouse; that New York, hitherto delinquent, had five hospitals in course of construction, which, with additions to old asylums then projected, would provide for all pauper insane; New Jersey had added to her hospital at Trenton, making provision for between five and six hundred, and was building one of equal capacity at Morris Plains. Maryland had hospital capacity enough. "In the hospital at Washington, planned by Dr. Nichols, the national government provides for about five hundred insane from the army, navy, and the district. All the Western States have, at least, one hospital; many of them more. One of the first things provided by the new States, after coming into the Union, has been a hospital for the insane. All the Southern States have at least one hospital, but they became so impoverished by the war that they are hardly able to maintain them, much less to meet the increasing demands for new ones."† On

* Now (1885) under the charge of Dr. Curwen.
† Letter written 1874.

the whole, therefore, there is an onward movement, and it looks as if public opinion, enlightened by the writings of American psychologists—especially the Annual Reports of the Hospitals for the Insane, and the manifestoes of the " Association "—would demand, and be willing to support the further extension of hospital accommodation which doubtless is called for, although it may be only gradually effected. "I can state two facts," writes Miss Dix to me (1876), concerning the state of communities in the United States, "an acknowledged obligation to *provide suitably* for *all* insane persons, whether chronic or recent cases—for the former *permanently*, for the latter, till cure is advanced, or recovery established. Much is said on the supposed rapid increase of insanity in the United States. I do not think this a sound proposition. Of course, the number of insane persons is vastly larger than ten years since, but the amazing increasè of population by a continually inflowing emigration from Europe, with the natural increase of native inhabitants, will create imperative need for a multiplication of hospitals for care and treatment."

I cannot conclude without a brief reference to an article which appeared in the "Lancet," Nov. 13, 1875, in which the writer brought very serious charges against the customary treatment of the inmates of the American asylums by their medical superintendents. I deeply regret that so unqualified an attack should have been made upon a body of honourable and humane men, and I am sure that the members of the Medico-Psychological Association in this country share in the regret. If the writer had spoken strongly in reference to the condition of the insane in workhouses and jails, or in *some* of the asylums, the case would have been entirely different; the language in regard to such is not too strong I dare say; but surely the only justification for a wholesale onslaught on the medical superintendents of the asylums in the United States would have been conclusive evidence of the alleged facts as a general rule in these institutions. On the contrary, Dr. Bucknill,* in regard to the several particulars specially mentioned by the "Lancet," entirely denies the correctness of the statements. To those familiar with the names, writings, or deeds of Butler, Earle, Gray, Nichols, Curwen, and others among living, and Brigham, Woodward, Bell, Ray, and Kirkbride, among dead medical psychologists, it sounds strange to read that " we are almost forced to the conclusion that our friends across the Atlantic have not yet mastered the fundamental principles of the remedial system." And stranger still (so far as regards the men of the first class), to hear that " they adhere to the old terrorism, tempered by petty tyranny." Can we be surprised that the feelings of men engaged in a noble and arduous work—the work of their lives—should be hurt when they read such charges made by members of the same profession? With what feelings, *mutatis mutandis*, should *we* read them? In the rejoinder made in the " Lancet " to Dr. Bucknill's letter, it is said, "We do not say *all* American asylums are bad." It is certainly to be regretted that this qualification, or rather a much larger and more generous one, was not made in the original article. Neither, on the other hand, do we say that all American asylums are good. We simply maintain that the sins of some asylum authorities—and these, as a rule, municipal rather than medical—should not be visited indiscriminately upon the whole body of medical superintendents of hospitals for the insane. An American physician, visiting St. Luke's subsequently to 1840, found chains in use. Had he in consequence stigmatized the English superintendents of asylums, as a body, as being in the habit of employing manacles, he would have committed a gross injustice, which they would have instantly resented. In the same way the American

---

* "Lancet," Feb. 12, 1876. Dr. Bucknill's impressions of American Asylums will be found in his little book, published by Macmillan in 1876, "Notes on Asylums for the Insane in America."

superintendents naturally feel aggrieved when a leading medical journal represents them, without (in the first instance) any exception whatever being made, as adhering to the old terrorism, &c.; resorting to contrivances of compulsion; using the shower-bath as a hideous torture; and, leaving their patients to the care of attendants, while they devote their own energies to beautifying their asylums.

Let us give credit where credit is due, and not involve in indiscriminate censure, worthy and unworthy superintendents, good and bad asylums; but if we denounce at all, let us confine our denunciation to those institutions in which ill-treatment is known to prevail.

The following is a list of the principal Hospitals for the Insane in the United States, with their dates of opening:—

1773. Williamsburg, Virginia.
1817. Frankford, Penn. (Society of Friends).
1818. McLean Asylum, Somerville, Mass.
1821. Bloomingdale, New York. (Hospital, 1797.)
1824. Hartford, Connecticut.
1824. Lexington, Kentucky.
1827. Columbia, South Carolina.
1828. Staunton, Virginia.
1832. Worcester (Chronic Insane), Mass.
1836. Brattleboro, Vermont.
1838. Columbus, Ohio.
1839. Boston Lunatic Hospital, Mass.
1839. New York City Lunatic Asylum (Women), Blackwell's Island.
1840. Augusta, Maine.
1840. Nashville, Tennessee.
1841. Philadelphia, Penn. (Hospital, 1752-1841.)
1842. Milledgeville, Georgia.
1842. Concord, New Hampshire.
1842. Mount Hope Retreat, Baltimore, Maryland.
1843. Utica, New York.
1845. Almshouse, Philadelphia, Penn.

Hospitals for the Insane.

1847. Indianapolis, Indiana.
1847. Providence, Rhode Island.
1848. Jacksonville, Illinois.
1848. Jackson, Louisiana.
1848. Trenton, New Jersey.
1851. Harrisburg, Penn.
1851. Fulton, Missouri.
1853. Taunton, Mass.
1853. Stockton, California.
1854. Hopkinsville, Kentucky.
1855. Dayton, Ohio.
1855. Jackson, Mississippi.
1855. Washington, District of Columbia.
1855. Flatbush, King's Co., Long Island, N.Y.
1855. Newburgh, Ohio.
1856. Raleigh, North Carolina.
1857. Austin, Texas.
1857. Dixmont, Penn.
1857. Northampton, Mass.
1859. Criminal Asylum, Auburn, New York.
1859. Kalamazoo, Michigan.
1859. Mendota, Wisconsin.
1860. Burn Brae, Kellyville, Penn.
1860. Longview, Carthage, Ohio.
1861. Tuscaloosa, Alabama.
1861. Mount Pleasant, Iowa.
1862. East Portland, Oregon (Closed 1883).
1866. Middletown, Connecticut.
1866. Osawatomie, Kansas.
1866. St. Peter, Minnesota.
1866. Weston, West Virginia.
1867. St. Louis, Missouri.
1867. Bellevue Place, Batavia, Illinois.
1868. Danville, Penn.
1869. Willard, Seneca Lake, New York.
1870. Cranston (State Farm), Rhode Island.
1870. Richmond, Virginia.

1871. Elgin, Illinois.
1871. Poughkeepsie, New York.
1871. Woodbridge, California (To Stockton in 1877).
1871. Lincoln, Nebraska.
1871. New York City Asylum (Men), Ward's Island, N.Y.
1872. Winnebago, Wisconsin.
1872. Catonsville, Baltimore, Maryland, (Hospital 1797).
1872. Jacksonville (Oak Lawn Retreat), Illinois.
1872. Athens, Ohio.
1873. Anna, Illinois.
1873. Independence, Iowa.
1873. Frankford, Kentucky, (now for feeble-minded children).
1873. Anchorage, Kentucky.
1874. St. Joseph's, Missouri.
1874. Middletown (Homœopathic), New York.
1875. Cincinnati Sanitarium, College Hill, Ohio.
1875. Napa, California.
1876. Morris Plains, New Jersey.

Hospitals opened after 1876.

1877. Worcester (New Hospital), Mass.
1878. Pontiac, Michigan.
1878. Danvers, Mass.
1879. Rochester, Minnesota.
1879. Topeka, Kansas.
1879. Kankakee, Illinois.
1879. Norristown, Pennsylvania.
1880. Warren, Pennsylvania.
1880. Goldsboro', North Carolina.
1880. Buffalo, New York.
1881. Binghampton, New York.
1882. Little Rock, Arkansas.
1883. Salem, Oregon.

Such is a brief sketch of the history of the insane in the United States in former years. The following pages will describe their present condition, so far as the author's visit in 1884 enabled him to judge of it.

CHAPTER III.

Present Condition of the Insane in the United States.

I. *General Management and Treatment.*

In my recent visit to the asylums of the States of New Hampshire, Vermont, Massachusetts, New York, Connecticut, Rhode Island, Pennsylvania, New Jersey, the District of Columbia, Illinois, Wisconsin, and Maryland, I found that, with some exceptions, their condition was satisfactory, many being admirably managed, and reflecting great credit upon all engaged in their administration. As a class, the American Asylum Superintendents are excellent men, devoted to their work, and as honourable, intelligent, and humane as those in any other country. I can, of course, speak only of the asylums and physicians I know. Judging from report, I believe there are institutions in some localities which are not in a very creditable state.* In fact, it is quite recently that the treatment of patients in one of the asylums in a Western State was admittedly most disgraceful. And in several of the institutions I visited, the rooms occupied by patients were quite unsuitable, and the amount of mechanical restraint indefensible. I may go a step further, and say that in regard to the latter point the number of

---

* The insufficiency of the provision for the insane is indicated by the statement made by an American physician, Dr. Dana, two years ago, that the condition of the insane in some of the Southern States "is particularly distressing;" and that in South Carolina hardly one-third can be cared for in the single State Hospital there. The asylum in Texas holds only one-fifth of the State insane. "The importunities of the few Medical Superintendents in the South show how negligent these States are." He asserts that the condition of the non-asylum insane in the Southern and Western States has little altered from what it was 10 years ago. "They are miserably kept, in jails, almshouses, and on poor farms, &c."

asylums is considerable, in which there is more resort to restraint than superintendents in England would approve, although I am by no means sure that in all these cases disapproval would be warranted. In many other American asylums there is either no restraint whatever, or it is so slight and manifestly necessary for surgical reasons, that hyper-criticism alone would find fault.

I have been favoured with certain unpublished returns of restraint made in 1880, but before giving them I must premise that it would be unfair to take these figures as giving a correct representation of the amount of restraint at the present time, because I am certain, from information received, and from my own observation, that the number restrained in asylums has within the last few years been greatly reduced. One is glad to know, also, that many who were then in almshouses, and were in restraint, have been removed to institutions for the insane in which little or no restraint is employed.

The number of patients in asylums in 1880 was 40,992, and the number reported to be under restraint 2,242, or 5·4 per cent. The mode of restraint was as follows:—

Camisole	887
Muff	526
Strap to bench	439
Handcuff	147
Ball and chain	21
Crib-bed	111
Form of restraint not stated	111
	2,242*

I should suppose the number returned under "crib-bed" is below the mark. Its incomplete return may be accounted for by the circumstance that it is scarcely regarded as mechanical restraint by some superintendents, inasmuch as it does not actually confine the limbs in one

---

* Also restraint by "personal attendance only," 1444.

position. Outside asylums a return is given of the use of crib-beds, but I wish now to restrict myself to asylums proper, not almshouses or private care. The "ball and chain" reported as in use in 21 cases in asylums sounds strange to our ears. There were none in any asylums I visited; probably they are to be found in some Southern or far Western institution. Returning to the crib-bed, which has caused so much acrimonious discussion, it would be disingenuous to deny that there are patients who are constantly getting out of bed, sometimes feeble elderly people; and others, restless, excited patients, who persist in standing up, and become very much exhausted; for whom it is an ingenious and sometimes effective device. I make this admission, as I do not believe in that definition of travellers which defines them as persons who go abroad in order to lie for their country; or, as the immoral Scotch proverb expresses it, "A travelled man hath leave to lie." At the same time, the crib-bed is to me an unpleasing object, and inevitably suggests, when occupied, that you are looking at an animal in a cage. Moreover, it is so temptingly facile a mode of restraint, and is on that account so certain to be abused, that I hope it will not be introduced into this country among the useful American inventions we are so glad to possess. That whatever its occasional utility may be, it may be abused, will be admitted when I say I counted 50 in use in a single asylum, and that a very good institution in most respects. At the celebrated Utica Asylum, under Dr. Gray, where a suicidal woman was preserved from harm by this wooden enclosure, my companion, Dr. Baker,* of the York Retreat, allowed himself to be shut up in one of these beds, but preferred not remaining there.

On examining the journal of the Bloomingdale Asylum, New York, of which Dr. Nichols is the Superintendent (there being 247 patients), I found that from the 1st of

* Dr. Baker and myself visited ten of the United States Asylums together. Including those in Canada, I inspected forty.

January, 1884, to the date of my visit, in the third week of October, 1884, two men had at times required restraint to prevent self-mutilation. I think no one will be disposed to criticise the resort to restraint in these cases. On the women's side of the house there had been no mechanical restraint for two years, and no seclusion during the year. I may add that during the several days I was at the Bloomingdale Asylum there was one individual, and only one, to whom it was necessary to apply mechanical restraint, namely, the doctor's collie, which it was needful to muzzle for a snapping propensity which suddenly developed!

I would here say that in visiting the American asylums I carefully refrained from making non-restraint the measure by which I estimated them, but looked rather at the general comfort of the house and the patients as a whole, and had regard to the evident character and intention of those in authority. I had, of course, a great deal of conversation with superintendents on this inflammatory topic, and I should say that they feel they have not always been fairly treated in the criticisms made upon them in this country.

What the American alienists complain of is this: they say that the British superintendents proclaim themselves the disciples of Hill and Conolly; that they protest in their writings against mechanical restraint as never necessary and always injurious; and yet when they (the Americans) visit our institutions they find some patients in restraint even in the best conducted asylums, often for the same reasons for which they themselves resort to it, and for which they have been so severely criticised. Moreover, in conversation with some of the superintendents of British asylums, they meet again and again, they say, with the frank admission, that the absolute disuse of mechanical restraint in accordance with the teachings of Hill and Conolly is not really held by them in theory, but is rather a "pious opinion."

Perhaps there has hardly been that public outspoken avowal of the occasional resort to restraint which the rare examples which do occur would seem to demand, whenever Conollyism is proclaimed as the adopted faith, and, by implication, the constant practice, of the medical superintendents of British asylums. Thus the difference between American and English practice (if not principle) is made to appear still greater than it really is. The intelligent and humane physician in America who declines to bind himself by any formula or vow, and who, in the exercise of his unshackled judgment, applies the camisole in extreme and exceptional cases, is practically at one with the English alienist who calls himself a disciple of Conolly.

Recurring to the allegation which led to these reflections, I will not deny that the American psychologists have some ground for their feeling on this question. It is not too much to say that an attempt has been made by some writers to divide cis-Atlantic and trans-Atlantic alienists into the sheep and the goats of the psychological kingdom. I shall rejoice if my reminiscences of American physicians engaged in the treatment of the insane have the effect of dispelling so exaggerated an opinion. As I have said, among the American alienists there is a large proportion of men as benevolent, intelligent, and devoted to their arduous work as their *confrères* in Britain; and while I sincerely hope that the practice of those among them who still resort to frequent restraint will ere long accord with the practice of some of their brethren who very rarely use it, I give them the fullest credit for pursuing their present practice with the honest conviction that it is the best for the patients under their charge.

Of *seclusion* I would say that while some superintendents, like Dr. Gray, state that they never resort to it, there was evidence of its use in many asylums to as great an extent as in Britain. In only one asylum, Danvers, Mass., (Dr. Goldsmith's) did I see a padded room.

With regard to the care and inspection of *suicidal cases*, there is no uniformity in the American asylums. It is rare to concentrate such cases in one associated dormitory. More often the attention of the night-watch and of non-suicidal patients in the same room is relied upon, and in very active cases mechanical restraint would no doubt be resorted to during the night. At the Middletown Asylum, Connecticut, superintended by Dr. Shew, the suicidal patients are usually placed in a large dormitory, and those who are very actively suicidal have also an attendant sitting up with them. I may mention that since this asylum has been opened, 17 years ago, 4,000 patients have passed through it, and there have been 14 suicides, five of which occurred in one year. At Utica, the suicidal patients are placed together.

Of 13,594 deaths occurring in American asylums in 146 years, 124, or nearly 1 per cent., were due to suicide. It is estimated that a suicide takes place in an asylum of average size every year and a half. Of admissions into asylums it is stated that from 15 to 25 per cent. have suicidal tendencies.

It is not usual to congregate all the *epileptic patients* together in the same dormitory. At the Government Hospital for the Insane at Washington, however, Dr. Godding has converted a large sewing-room into an associated dormitory for this purpose, and a large associated dormitory has for some time been in use at Kankakee.

With regard to the strictly *medical* treatment of the insane, I do not think there is much, if any, difference between the American and the English practice. Perhaps fewer new drugs are administered in the former than the latter. Hyoscyamine (Merck's) is used hypodermically (one-tenth to one-twelfth of a grain) in a good many asylums. Chloral and bromide of potassium are given to much the same extent as they are in England—if anything, more in America—20 grains of each being a frequent dose at bed-time. In an asylum containing 780 patients, 15 draughts of chloral were prescribed on the evening of the day of my visit. At the Butler Hospital, Providence (R.I.), Dr. Sawyer was giving 10 grains of chloral, with 15 of bromide, at bed-time, repeated in the night if necessary. The usual remedies for intercurrent physical disorders are, of course, resorted to—quinine, iron, &c.—but nowhere did I hear of any special remedy put forward with confidence for the treatment of mental disorders. I am afraid that we have nothing either to teach or learn from each other in the therapeutics of insanity. The American asylums are well supplied with baths, which are employed to much the same extent and under the same circumstances as with us, but I did not hear of the prolonged bath being used in any asylum. Shower baths are, I believe, never made use of, and I am not aware that any asylum in the States, besides Kankakee and Utica, is provided with the Turkish bath. I did not meet with the wet or dry pack. At the annual meeting of our Medico-Psychological Association, 1884, Dr. Nichols, the Superintendent of the Bloomingdale Asylum, New York, in his able and interesting speech, stated that he employed the warm bath, with cold water to the head, in suitable cases, followed by rubbing the whole surface of the body with whiskey as a swelling is rubbed with liniment, and that this treatment often succeeded better in inducing sleep than did the administration of any drug ; at the same time allaying the fever and saving the strength of feeble patients. I observed the truth of his statement that shaving the head and applying counter-irritation do not form part of the practice of American asylums, this treatment being regarded as of doubtful advantage, and therefore scarcely justifiable. Preference is therefore given to blistering or cupping the nape of the neck, the temples, or behind the ears. Dr. Nichols relies on opium in a few cases of mania and in some of melancholia. When I was at Bloomingdale (297 patients) I examined the Medical Record, and found that at the time of my visit no patients were taking hypnotics or sedatives, while 34 were taking tonics, 52 "miscellaneous medicines," and 7 had medical baths. I do not think that the open-air treatment of maniacal patients is so frequent in the American asylums as in our own. It is not so common a practice there as here to employ one or two attendants to take such a patient out into the grounds or airing court, and allow him to work off his excitement by exercise, as well as improve his health by plenty of fresh air. Special attention has been paid to uterine affections in their relation to insanity in one at least of the American asylums. This has been facilitated, I should say, by the appointment of lady physicians in several of their institutions. Of two of these, Dr. Margaret Cleaves and Dr. Garver, the former recently, and the latter now, the physician on the female side of the Harrisburg Asylum, and both intelligent and very anxious to advance the therapeutics of uterine insanity, the experience has been that some cases have been benefited by the treatment pursued, consequent on the knowledge of uterine disorders obtained by examinations which probably would not have been made in the ordinary routine of asylum treatment. At the same time I am afraid it must be confessed that these results are but scanty, and fall far short of what had been anticipated from the particular attention thus paid to this department of practice, under, as I consider, very favourable auspices.

Useful, then, as this treatment has doubtless been, it is very far indeed from justifying the opinion of those obstetric physicians who think that if the superintendents of asylums would only examine and treat the uterine condition of many more patients than they are wont to do, the number relieved or cured would be much greater than it is at the present time.

In connection with treatment I would briefly refer to the *Recoveries* and *Deaths* in the American Asylums. I have not, however, a sufficient number of trustworthy statistics from which to draw any safe inference. So far as I have obtained returns, I find that the rate of recoveries to admissions varies from 20 to 42 per cent. It is to be regretted that so many of the reports of these institutions give only the statistics of the past year, and not the results since the opening of the institution. As is well known, Dr. Pliny Earle has carefully studied the curability of insanity, and has done good service by pointing out the fallacy of counting all the recoveries of a single case as if they represented so many recovered insane persons. He has shewn that in the American asylums the recoveries as reported were more frequent in the early than in the later periods of their history—a result which he attributes to the larger number of chronic cases transferred from almshouses to asylums; to the increase of general paralysis; to different modes of calculating recoveries; to the greater care exercised not to return patients as "recovered" who have only improved; to the small number of cases upon which early statistics were based; and lastly, to the greater tendency to retain patients under care, to save them from relapse. As regards mortality, in the best asylums it does not exceed 5 to 7 per cent., calculated on the average number resident.

I must here say a word on the alleged difference of type in *the form or intensity of mental disorders* in the two countries.

It has seemed to several American alienists, on visiting our asylums, that English madmen present a milder type of excitement than those of their own country. Dr. Draper, the very intelligent superintendent of the Brattleboro Asylum, Vt., on visiting this country some years since, was strongly impressed with the difference between American and English patients, and suspected that violent excitement was comparatively transitory in our asylums, while it continues for months and even years in theirs. It is difficult, I think, to form an opinion on the subject by visiting asylums for a short period in either country. I attach great weight to the testimony which I obtained from English attendants in American asylums, who had previously been employed for years in this country. I questioned them closely, and they were, I found, distinctly of opinion that the patients they had had to do with in England were more violent and difficult to manage than those in America. The latter, according to my informants, are not (contrary to what might have been supposed) more impatient of control, or more tenacious in maintaining their individual rights.

As to the relative frequency of *general paralysis*—there has hitherto been, and there still is, decidedly less in the American asylums than in England, but I am sorry to say it is clearly on the increase, and bids fair to equal in amount that witnessed in the mother country. Till recently it was very rare indeed in women, but is becoming more frequent.

A few words may here be said in regard to the vitally important subject of *Employment*. I found at the Utica Asylum, containing 600 patients, that the average percentage of men employed is about 35 per cent., and of women nearly 38 per cent. I should add that from one-fifth to one-fourth of the inmates are pay patients.

At the Willard Asylum (N.Y.), out of 1,758 patients 801 are reported to be able and willing to engage in some kind of occupation.

Again, at the new asylum at Worcester, Massachusetts, where there are 780 patients, 38 per cent., namely 102 men and 218 women, were occupied.

At the Northampton Asylum, the laundry work for an average of 530 persons is done with only two assistants (women), whose aggregate wage is £7 a month. For the last fifteen years, nearly or quite three-fourths (Dr. Earle thinks it may safely be placed at two-thirds) of all the *necessary manual labour* upon the premises have been performed by the patients. With an average number of over

30 cows, the milking is all done by patients—an employé over-seeing them. The poultry-house is under the sole charge of a patient. And, as closely connected with employment, may be mentioned the remarkable extent to which, at this institution, in-door recreation in some form is carried out, namely every evening during the year.

It is creditable to the Norristown Asylum (Penn.) that of 510 male patients 167 were employed out of doors. Sixty-four were engaged in the brush-shop alone.

At the New York City Asylum for males (Ward's Island), under Dr. MacDonald, a printing press is in constant use, and there were four patients at work on the day of my visit, bookbinding being done as well as printing. Of the total number of patients (1,494) 34 per cent. were employed, 219 in outside and 298 in inside work. As many as 1,154 go out for exercise. Although I can hardly adduce bathing as an example of work, I may mention, as a praiseworthy attempt to occupy the patients, that salt water has been brought into the grounds so as to form a large bath 220 feet long by 30 feet broad, and from $4\frac{1}{2}$ to $5\frac{1}{2}$ in depth, where a number of patients bathe in the open air to their heart's content. From 500 to 600 patients bathe every day. There were as many as 200 in the water when I was there, and they evidently enjoyed their immersion immensely. There is a large shed under which they dry themselves. This is the second year the bath has been in operation, and it has proved a great success.

As a useful mode in which to employ patients I may here mention that, at the Pennsylvania Hospital for the insane, till recently under Dr. Kirkbride's charge, a considerable number are engaged in pottery work, the clay being moulded by them into various useful and ornamental forms, specimens of which are collected in a room in the institution.

At the Middletown Asylum, Connecticut, superintended by Dr. Shew, 45 per cent. of the 892 patients were employed, 26 being engaged on the farm, 37 on the grounds, and 18 in the stables.

When I come to speak of the system of providing for the chronic insane in the State of Wisconsin, it will be shown that a very considerable amount of work is done by them in the small county asylums.

It must be admitted that the above percentages are somewhat low, even for mixed asylums (as some of them are) and contrast strongly with our Sussex asylum, where 66 per cent. of the men are employed, exclusive of ward cleaners, 116 being engaged on the land; or with Brookwood (Surrey), where, of 318 men, 163 are engaged on the farm and garden, and 66 at various trades.

There are two reasons assigned for this difference. The one is the greater independence of the people in America, the other the character of the climate. When I observed one day to an American doctor that, on such a sunny day as that, our patients would be nearly all out of doors, "I should think they would in *England*," he replied, ironically, for he knew something of this sunless island of ours, and was not surprised to hear that when the sun does shine everyone turns out to see it. But, joking apart, I should hold that the English are in advance of the Americans in this very important matter of out-of-door occupation; and this is the more to be regretted when one considers the liberal supply of land which is attached to many State asylums. There is already a movement in this direction, and I am inclined to think it will be found that neither independence of character nor peculiarities of climate will ultimately prevent the system being carried out to a greater extent than it is at present.

Dr. Dana refers, in an article in the "Journal of Nervous and Mental Disease,"* to the lack of sufficient work and amusement as an evil still existing in a large number of the asylums in his country, but adds that in some States the asylums are so miserably provided with money and

* Vol. ix, No. 2, April, 1882.

even grounds that really little can be done. Among other State asylums, those of New York and Maryland appeal to the authorities for power to give their patients more work, exercise, and amusement.

It must, in this connection, be remembered, in justice to American asylums when compared with our own, that while in England men can work out of doors during the greater part of the year, there is little or nothing to be done in many of the American States from November to April.

Very much is certainly done in the way of providing instructive amusements for the patients. I have already referred to the systematic manner to which this is carried out at Northampton. Dr. Curwen, at the Warren Hospital for the insane, in common with many other superintendents, keeps up frequent magic lantern exhibitions during the week, taking a certain number of pictures and explaining them each evening. He informs me that he has imported from London more than 500 photographic views in Great Britain and Ireland. These readily afford an opportunity for observations on points of history, biography, and geography. One evening is given to natural history, with photographs of animals, birds, structure of plants, &c. The assistant medical officer takes one night for music or theatricals, and on the sixth night Dr. Curwen himself reads on various subjects, poetry, &c. In this way every evening in the winter has something to occupy the patients. Dr. Curwen thinks that too little attention, as a rule, has been given to the matter of instruction and diversion, which, while they involve much time and labour, prove of corresponding benefit to the patients.

II. *Lunacy Legislation.*

I now pass on to the question of Lunacy Legislation, and as this excites so much interest in England at the present time, I will mention the practice pursued in several of the American States, the result of laws which have been enacted for the protection of the insane. As these laws vary with every State (nearly 50), it is impossible to do more than single out three or four as examples of the different plans which have been adopted, including those of the most drastic kind. It is as they bear upon admission into asylums, involving, as this does, the deprivation of liberty, and the inspection of asylums, that we are most concerned with these enactments, and the practice pursued.

A. *Admission.*—(*a.*) I will first take the simplest possible form of admission as it is found in the State of *Connecticut*. The request for admission of a private patient, signed by a guardian, near relative, or friend, simply desires that A. B., of C. D., may be admitted as a patient into the hospital for the insane.

The certificate of one physician is sufficient, and runs thus :—

"I hereby certify that I have, within one week of this date, made personal examination of A. B., of C. D., and believe him to be insane."

This is subscribed and sworn to by the physician before an officer authorized to administer oaths, who certifies to the respectability of the physician and the genuineness of the signature.

(*b.*) If we now go to *Pennsylvania*, we find a considerable advance made in the stringency of the checks on improper admission into asylums, although I should observe that the law of Pennsylvania allows persons to place themselves voluntarily in an asylum for a period not exceeding seven days, on signing an agreement giving authority to detain them, which may be renewed from time to time for the same period.

In ordinary cases, in accordance with an Act of 1883, passed in consequence of a Commission appointed by the Governor of the State to report upon the Lunacy Laws, it is necessary that the medical certificate should be signed by at least two physicians in actual practice for five years, who shall certify that they have separately examined the patient, and believe him to be insane and requiring the care of an asylum. They must not be related by blood or marriage to the patient, nor connected medically or otherwise with the institution. This certificate must be made within one week after the examination of the patient, and within two weeks of his admission. Further, it must be sworn to or affirmed before a judge or a magistrate, who must certify the genuineness of the signature, and the standing and good repute of the signers. It is not, however, necessary that the judge or magistrate should examine the patient, or express any opinion in regard to his insanity.

The order and statement are signed by the person at whose instance the patient is received. The statement comprises the chief points of importance, but is not so full as our own.

Copies of the admission-papers are forwarded to the Committee on Lunacy, which is a section of the Board of Charities, within seven days of admission.

(c.) I will now give an example of a more stringent procedure, and for this purpose draw my illustration from the State of *Massachusetts*. Here, not only are the certificates of two physicians required (the facts upon which their opinions are founded being specified), but the Lunacy Act of this State requires that no person shall be committed to a lunatic hospital, asylum, or other receptacle for the insane, public or private, without an order signed by a judge of one of the various courts enumerated, certifying that *he finds that the person committed is insane, and fit for treatment in an asylum.* Some discretionary power is, however, permitted, for if the judge thinks it undesirable to see the patient, he may certify to that effect, and still commit him. Again, if he is in doubt, he may summon a jury of six to his aid.

To obviate the difficulties which might arise from requiring a judge's order and examination in urgent cases (difficulties to which we are just now so much alive), an emergency certificate is allowed, upon which the Superintendent may receive and detain a patient for five days. This document, signed by two physicians, certifies that the patient is labouring under violent and dangerous insanity, and it is accompanied by an application for admission from the Mayor or one of the Aldermen of the place in which the patient resides. The permitting this exceptional action in cases of emergency must certainly materially lessen the inconvenience of the Massachusetts Act.

(*d.*) The last example, and the most stringent of all which I have to give, is in operation in the State of *Illinois*, where the law requires that no one shall be deprived of his liberty by being placed in an asylum without trial by jury. Instead of giving the details of the Act which requires this proceeding, I will briefly describe what I myself saw of trial by jury of the insane in Illinois, when I was at Chicago last October. I went, accompanied by a solicitor (Mr. McCagg), on what is called "insane Thursday," to the County Court, where the trials are held, presided over by Judge Prendergast. Below the Court were some rooms occupied by insane persons awaiting their trial. In the Court there were about forty spectators. In one corner of the room sat the jury of six, the foreman being a doctor. When the patient was brought in for trial, a physician (Dr. Bluthardt) employed by the Court, gave the result of his examination. A friend of the patient also gave evidence, and the judge asked a few questions. The jury then retired into an adjoining room to consider their verdict, and another case was tried. The consideration of each case did not occupy much time, but there was no unseemly haste. One patient was too acutely maniacal to be examined in the Court, out of which he was quickly conveyed, restrained by a leathern muff, into another room, where the jury and the official doctor went and examined him. I followed. The jury, very properly, made short work of the case. In one instance, after a careful inquiry into all the circumstances, the man was not found insane, although he was evidently not quite right, and it was agreed that he should go and reside with a farmer who was a friend of his. I had an opportunity of conversing with the judge, who told me that he regarded the law under which these trials are conducted as quite satisfactory. "Insane Thursday" is likely, I was told by others, to remain an institution in Illinois, as popular feeling demands trial by jury as a right. The publicity, however, is a serious objection; and I was informed that people often keep their friends at home rather than make their insanity known. The circumstance is almost sure to come out, although they often bribe the newspaper reporters not to report their friend's case, and though the judge, as he told me, sometimes considerately defers the trial of those cases in which he knows there is a dread of publicity until all the others have been disposed of, and the reporters have left the room.

In regard to jury-trial of the insane, Dr. McFarland, of Jacksonville, Ill., thus writes to another American physician, Dr. Parsons: "The Illinois law of which you inquire is injurious, odious, barbarous, damnable, and you may add as many more expletives to it as you please, and still not say the truth in regard to its evils. . . . Every superintendent of an asylum in the State is most eloquently pleading for a change in this detestable system; the Board of State Charities urges the change most forcibly; a Bill is before the Legislature, reported favourably upon; the Chairman of the Judiciary Committee is a true champion of the reform; but all, as I fear, will amount to nothing, because there are a few fanatics who raise the hue and cry over an imaginary bugbear."

Dr. Parsons himself objects to the jury-trial of the insane on the grounds that as the removal from home to an asylum necessitates publicity, to an odious extent, hospital treatment must be delayed in many cases until prospects of recovery have been seriously lessened; that the transfer to a Court and the incidents of the trial often endanger the life of a patient; and that, in not a few instances, patients become dangerously excited by having to appear in Court as defendants.

Dr. Jewell, of Chicago, highly disapproves of the practice; as does also Mr. F. H. Wines, the Secretary of the Illinois Board of Public Charity. Mr. Wines writes: "A delicate woman, for example a case of puerperal insanity, is dragged from her bed in winter across the country to the county court, and carried into the Court-room, more dead than alive, before she can be taken to the hospital. . . . The effect of the trial on the patient is often terrible. He is impressed with the conviction that he has committed some crime, he knows not what; he believes himself to be consigned to a prison; possibly he has a sense of having been dealt with unjustly and foully wronged; he looks upon the officers of the hospital as conspirators in a plot; it is long before his suspicion of them can be removed."

Dr. Dewey, of Kankakee, also objects to this law. He says that the unfortunate influence exerted by the patient's having been treated as a criminal "is consequently apparent among the patients in our hospitals, an undue proportion of whom are impressed with the idea that they are unjustly accused of some crime, and tyrannically held in confinement. . . . The best feelings of all right-minded persons are outraged by seeing presented in Court the depraved and unnatural acts and speech of otherwise reputable men and women."

From the description now given of these well-intentioned, if mistaken, enactments in force in the United States, it will be seen how much care has been taken in some of them to guard the liberty of the subject. There is certainly a good opportunity afforded of trying and observing various experiments in lunacy legislation, differing in their degrees of stringency, but all having for their object the protection of those who are not able to protect themselves, and whom we all wish to see guarded from anything like unjustifiable interference, so long as this can be done without disadvantage to themselves, disaster to their friends, or danger to society. England would do well to profit by these experiments. She would hardly be induced, I should think, to copy the law of Illinois, and demand trial by jury for every alleged lunatic, private or public. I am not aware that in Massachusetts any considerable evil * has attended the carrying out of the law requiring the examination of the patient, whether private or public, by a judge, guarded as this law is by an emergency certificate. Little, if any, objection applies to the simple guarantee on the part of the judge of the respectability of the signers of certificates, so long as this is accompanied by an emergency clause, by which the danger of delay in the admission of acute cases is obviated. It is worthy of consideration whether that part of the Pennsylvania law, requiring the physicians who sign the certificates to have been five years in practice, is not worthy of adoption. Whether this, or the guarantee of the judge in regard to his respectability, would meet the demand now being made in this country

* It is an evil, however, that this procedure frequently necessitates the detention of patients in gaols over night, in order that the judge may see them. Notwithstanding this objection Dr. Earle writes to me—" In my opinion the existing law in Massachusetts in regard to the commitment of lunatics is none too stringent. It occasionally seems somewhat burdensome to the friends of the patient; but that is a small consideration when compared with the defects in former laws which it has repaired, and the deficiencies it has supplied. It not only furnishes far greater safe-guards against improper commitments, and consequently a more sure protection to the rights of the individual, but judging from my experience, it is a great relief to the Superintendents of the Hospitals from whom it has removed all responsibility, in *all* cases, for the commitment of the patient.

" The emergency section of the Mass. law has been frequently resorted to in Boston, but here (Northampton) we have had only two or three cases during the nearly five years since the Act took effect. I have myself heard no complaint of the law arising from its requirement of the examination of the patient by the judge, but such a complaint would be much more likely to occur in the large cities than in a community like ours, where the population is very largely rural."

for greater checks upon the admission of patients into asylums, is another question; for the public, or a certain section of it, might feel very suspicious in regard to mental physicians whom no judge or magistrate would for a moment hesitate to guarantee.

B. *Inspection.*—I now turn to the other point of Lunacy Legislation to which I have referred, namely, the vastly important subject of inspection.

(*a.*) At the present time the inspection of asylums is usually performed by State Boards of Charities, which have been constituted in a considerable number of the States, and are likely to increase. They had not been established in three of the States I visited, New Hampshire, Vermont, and Maryland. In Illinois and Wisconsin, as we shall see, they have been very active bodies. In Massachusetts there is an active and effective Board, entitled the State Board of Health, Lunacy, and Charity, established in 1879, when two previously existing Boards, the State Board of Health and the Board of State Charities, were abolished, and their functions merged into the present Board. The members of the Board, with the exception of the Inspector and clerks, receive no compensation whatever for their services, the expenses incurred in travelling being, of course, re-imbursed. Whatever may be the reforms which it is desirable to introduce from America into England, I suppose this is hardly one which will find favour; at any rate, it is not likely to receive support from the Board in Whitehall Place. I was fortunate in meeting with the Inspector, Hon. F. B. Sanborn, a highly intelligent man, interested in all questions connected with lunacy, and liked by the superintendents whose asylums he inspects. He was for several years Secretary of the Board of State Charities, and has acted as Secretary to the present Board, which at the present time has no official Secretary, but a "Clerk and Auditor." The Board consists of nine persons, who are appointed by the Governor of the State, with the advice and consent of the

Council. It has the general supervision of the State Lunatic Hospitals and the State Almshouses as well as Schools. It must visit at least once a year all places where State-paupers are supported, and must inspect every private asylum or receptacle for the insane at least once in every six months. It is expressly enacted that the Board shall act as "Commissioners of Lunacy," with power to investigate the question of the insanity and condition of any person committed to any public or private asylum, or restrained of his liberty by reason of alleged insanity in any place within the Commonwealth, power being given to the Commissioners to discharge any person not insane, or who can be cared for, if discharged, without danger to others or injury to himself (Act of 1882, chapter 87, section 1).

Dr. Pliny Earle (Northampton, Massachusetts) spoke favourably of the working of the Board. Others evidently felt averse to a Board composed chiefly of laymen and laywomen, and were inclined to resent interference. That this feeling is natural must be admitted. Some irritation and annoyance will almost inevitably arise at times, in regard to advice tendered on points upon which doctors ought to be, if they are not, better judges than their advisers. But it appears to me that, officious and harassing as individual members may sometimes be, such a Board is of use, and must certainly be continued until medical Lunacy Boards are introduced; and even then I should regard it as very desirable to secure the unpaid services of the same class of men and women as visitors, though no longer Commissioners. They would make any suggestions which might occur to them to the Trustees and to the Commissioners in Lunacy. Ladies may prove invaluable in this way, for they often see the necessity of certain comforts and changes which may escape the attention of officials, and which, although seemingly very small, add greatly to the comfort of the patients.

(*b.*) In Pennsylvania a Committee of the State Board

of Charities has been established as a Lunacy Board, whose business it is to attend to the lunacy department of the Board only. Formerly there was a mixed Committee, but it has been found best to separate the duties of asylum visitation from those of other charities. Only the Secretary, Dr. Ourt, is paid, and his time is occupied in the inspection of the asylums of the State, the condition of which he reports to the Committee, whose members visit according as the necessity arises. One able member, Dr. Morton, is the son-in-law of the late Dr. Kirkbride, and son of the author of "Crania Americana." With him I visited Norristown, the excellent asylum near Philadelphia, under Dr. Chase and Dr. Alice Bennett, which has contributed largely to the practical solution of the question of providing accommodation for the chronic insane in Pennsylvania.

(c.) In the State of New York there exists, in addition to the State Board of Charities, a Lunacy Commissioner, Dr. Stephen Smith, to whom, as well as to Dr. Gray, I am under obligations for help rendered in forwarding the objects I had in view. I have reason to believe that his visitations and reports have done good service. I regret that the salary of this office is not such as would allow of a physician renouncing private practice. It is no doubt considered that his whole time could not be occupied in inspecting the State-institutions in which the insane are confined.

This is one of the difficulties connected with the appointment, in each State, of Lunacy Boards on our model —a proposal which has been much discussed by asylum superintendents in America.

I found from Dr. Stearns, the Medical Superintendent of the Hartford Retreat, Connecticut (so long associated with the name of Dr. Butler), which I visited with great satisfaction, that he was disposed to regard the appointment of a Lunacy Board with favour, if it could be administered under the same conditions as in Britain. He fears it is impracticable in his own country, because there is no central appointing power or authority. Each State must make its own appointment; but the number of asylums, unless it be in the State of New York, is too small to make it worth while or possible to set apart one Board for such a purpose. Dr. Stearns is well acquainted with Britain, and fully concedes the benefit which has accrued from the appointment of our Lunacy Boards. A stranger may well hesitate to suggest a plan which a man so impartial and well able to judge as Dr. Stearns regards as not feasible in America. It seems a pity that some States cannot join together, and so make it possible to have men, with ample knowledge of the insane, constituting a Board of Inspection without other duties to perform. But I am informed that such a scheme is not in accordance with precedent, and would never be adopted.

Again, Dr. Shew, the Superintendent (as already mentioned) of the State Asylum, at Middletown, has united with Dr. Stearns in obtaining information about Lunacy Boards, which has resulted in a report upon this subject. While the general sentiment expressed in this report is in accordance with what I have just stated, they heartily approve of a Supervising Board of some kind in each State of the Union, and they propose that such Board shall consist of at least five members, eminent in psychological or humanitarian work. They wisely do not contend for a mere name, whether "Commissioners in Lunacy," "Inspectors of Charities," or a "Board of State Charities," provided that the Lunacy Board in England or Scotland be taken as, on the whole, the best model.

III. *Provision for Chronic Insane. Segregation.*

The subject of the best form of provision that can be made for the chronic insane has, during recent years, greatly occupied and agitated the minds of physicians, philanthropists, and the Legislature, in the United States. It will tend to a clearer view of the claims made upon the

States of America in connection with their insane population if I state that at the last census in 1880, the number reported to be insane amounted to 91,959, or 1 in 545 of the general population, and the number of idiots to 76,895, making a total of 168,854. The distribution of the insane was as follows :—

In hospitals and asylums for the insane*	40,942	or 44·42 p.c.
In other institutions...	235	or ·26 p.c.
In almshouses	9,302	or 10·12 p.c.
In jails, &c. ...	397	or ·48 p.c.
At home or in private families	41,083	or 44·68 p.c.
Total	91,959	

Thus, no fewer than 50,782, or 55·28 per cent., were not in any special institutions. It is calculated that the annual increment in the numbers of the insane in the States is about 5 per cent. The asylums have been much crowded; and many efforts have been made, and some carried out, to supplement existing asylums by buildings of a less expensive character. Of this we witness one example in an Eastern and another in a Western State, namely, the Asylums of Willard† and Kankakee, which are examples on a large scale of what has been and is being done in the same direction by other institutions. This problem, which has occasioned so much discussion and led to such definite action in England, was discussed at a meeting of the New England Psychological Society in Boston last September, when I listened with much interest to the various opinions expressed by its members, elicited by a paper read by Dr. Quinby, the Superintendent of the Asylum for the Chronic Insane at Worcester, Mass. Considerable diversity of sentiment prevailed. Some preferred retaining the acute and chronic in the same building, the necessary additions being made as accumulation rendered the original capacity insufficient; others

---

* It is said that 90 per cent. of the insane in this category are incurable.
† So called from Dr. Willard. It is at Ovid, Seneca Lake, N.Y.

advocated distinct asylums, whether in the same or different localities. All admitted the existing pressure, and the necessity for further accommodation. The venerable President, Dr. Pliny Earle, observed that he did not see what disadvantage would result were he, for example, to reside in one building and an assistant medical officer in another, one taking charge of the acute, the other of the chronic cases. On the whole the balance of opinion in this debate was in favour of annexes or cottages on the same estate as the original hospitals.

I must here dilate a little on the Willard institution. Its establishment was the outcome of an attempt made in the State of New York to grapple with this problem of chronic insanity, and to rescue lunatics from the neglect and cruelty from which many of them suffered in almshouses. In common with some similar attempts, it was a considerable departure from beaten tracks and the tradition of the elders. Dr. Chapin led the way, and superintended the group of buildings known as the Willard Asylum, from its opening in 1869 to 1884. He informed me that the idea was suggested to him by the Fitz-James Colony, Clermont (France). I met Dr. Chapin in Philadelphia, where he has succeeded the lamented Dr. Kirkbride in the management of the Pennsylvania Hospital for the Insane; and I conversed with him very fully in regard to the disposition and arrangement of the detached buildings at the Willard Asylum, and heard from him in what respects the experiment has proved a success. Although I regretted being unable to visit the institution itself, I was made so thoroughly acquainted with it by the description of my informant that, like a certain royal personage who seriously maintained before Wellington that he was present at the battle of Waterloo, I have for some time entertained the idea of having actually visited the Willard Asylum. Be this as it may, it is always better to behold the *genius loci* than merely the place from which it has departed; that is to say, when one cannot see both.

When "Willard" was opened, there were as many as 1,500 lunatics in almshouses in the State of New York, while there were only 500 in the State Asylum at Utica, under the charge of Dr. Gray. Four years before Willard was opened, there were as many as 200 pauper lunatics in chains in the almshouses of the State.

Dr. Chapin claims as important advantages in the Willard Asylum system: first, economy of construction; second, economy of maintenance; and, third, greater facility for taking patients out to work on the farm.

Without entering into details, I may state that it provides for about 1,800 patients, and is the largest asylum in the States. The main building contains nearly 600 patients; Dr. Chapin, however, disapproves of so large a number, preferring 300. There are also twenty detached blocks, in four groups, each group comprising five. There is another building, formerly an agricultural college, adapted to the requirements of the insane. The cost of three buildings, which are to be erected for 200 patients in all, will be under £50 per bed, exclusive of land and furniture. One of the detached buildings accommodates 250 women, a matron and an assistant physician residing there. All patients, when admitted, go in the first instance to the central building, and are subsequently classified and distributed according to their mental condition. Profiting by his experience at Willard, Dr. Chapin would prefer having only 50 patients in a detached building. At present there is a kitchen provided in every house, but if he only had houses with a capacity for 50 patients, Dr. Chapin would plan to have, in a separate building, a kitchen and a large dining-room for common use.

I now ask the reader to accompany me to the other locality, Kankakee, Illinois, some 50 miles from Chicago, where a similar, though in some respects a different, experiment has been made. Some years ago a gentleman was deputed by this State to visit the asylums of Britain, before a decision was arrived at as to the best mode of construction to adopt in the erection of a new State Asylum. Those who, like myself, met Mr. Wines in England, will remember the careful and intelligent study which he made of our institutions. On his return he recommended the erection of a main building and several others entirely detached from it, with the view of avoiding one huge many-storeyed building, and facilitating classification by providing dwellings as much as possible resembling those to which the patients had been accustomed at home. Mr. Wines, as Secretary to the Board of Public Charities, has made this institution his special care, and continues to take a lively interest in it. Segregation is here carried out to its fullest extent, and, as some think, to an objectionable extreme. Apart from any errors of detail which may have been committed, it is a pleasant thing to see this breaking up of buildings on so extensive a scale. It must do good. The air blows more freely and freshly through this group of houses, which will soon form a little village, than it would or could through monster structures filled from top to bottom with the insane. I saw with pleasure, one evening, a number of patients sitting at ease under the verandah of one of these cottages, some of them, if not all, having been engaged in wholesome work on the farm during the day. There was an air of freedom and homeishness which is necessarily more or less lost in an ordinary asylum, especially when of giant proportions. And I may say that, although not a few in the States look with a critical eye at what they regard as segregation run mad, the principle itself, that of providing many small buildings in place of a large one, is rapidly advancing in the United States. In fact, there has always been a strong feeling against the aggregation of a large number of patients under one roof; but until comparatively recent times it did not become necessary (or perhaps it would be more correct to say, the necessity was not recognised) to provide in every State of the Union asylum accommodation for a large number of acute and chronic cases.

The asylum of Kankakee was built in 1878 and '81. When I visited it I found some of the detached buildings in course of erection, and more will eventually be built. There are about 480 acres. Dr. Dewey, the superintendent, speaks warmly of the success of the experiment. The number of patients at the present time is 615 (370 men, 245 women). Eventually there will be accommodation for 1,500 patients. One large unfurnished building contains a dining hall for 500 patients, 85ft. by 67ft., and is very light and well proportioned. Below this room is a kitchen. The patients will eventually come here to dine from some of the detached buildings for males.

At the present time, dinners are conveyed by hand-cars from the main building to the cottages, and, as the distance is considerable, there has been a good deal of criticism on this part of the arrangement, but I was assured that the food was not more chilled in the transit from the central kitchen to the detached buildings than when conveyed to the extreme ends of the main building. It is intended, however, to make use of a special apparatus invented by the New York Catering Company, for keeping meat warm when carried from one spot to another.*

I may add that two of the buildings are infirmaries, one for each sex; the rooms are well adapted for the wards of a general hospital. I regret to say that the male infirmary, which was only just occupied at the time of my visit, has

---

* This, I hear from Dr. Dewey, has been tried and has failed to give satisfaction. Food-cars of galvanized iron have now been introduced that are closed tightly, and surrounded with an air chamber, through which a small stove under the car circulates a continuous current of hot air. These prove entirely successful. They run easily by hand over the concrete walks which have been provided. There is no covered corridor required. With the tea and coffee made in each dining-room, the warming closets provided for each, and the tightly closed car, Dr. Dewey maintains that they are able to get food to the dining-rooms in as good condition as would be possible in any linear hospital of equal or even much less extent. It is admitted that, in those rare winters when the snow-falls are deep and frequent, there are disadvantages, but it is held that the advantages obtained by the detached system compensate fully for such drawbacks. See Chap. iv.

been recently burnt, the lives of a number of the patients being lost. There are two houses called "Convalescent Homes," one for each sex. Another is the "Relief Building," which provides for 50 epileptic patients and 21 criminal lunatics. They appeared to be under very good care (one attendant was from an English asylum), and no patients were under mechanical restraint. There is also a recreation hall, used as a chapel as well, which accommodates 350. As is usual in American asylums, ministers of various denominations from the neighbouring town conduct the services in rotation. While I was at Kankakee there was an evening entertainment in this hall, at which there was music and dancing. In harmony with the ruling idea of making as little difference as possible between home and asylum life, the recreation hall is not connected by any covered passage with the main building; for it is thought that while, on very cold evenings, some patients may be deprived of the pleasure of the entertainment, something is gained from the freshness and change consequent upon having to dress and go to a distinct building.

I should mention that at Washington also, Dr. Nichols and his successor Dr. Godding have favoured the erection of small additional buildings at Saint Elizabeth, the Government Hospital for the Insane. There was, in the first instance, a cottage for the patients of colour; and subsequently, detached buildings have been erected for (1) feeble-minded children; (2) quiet, working patients, "The Home;" (3) the "Relief House," for tranquil patients; and lastly, "The Rest," the name fitly given to the *post mortem* room and mortuary. Here the body is placed in a refrigerator, which is situated between these two rooms, and can be conveyed to either when required.

Dr. Shew has carried out the plan of separate buildings at the Middletown Asylum, Connecticut, where there are 900 patients. Here is one annexe for 325 men and women patients, the cooking being done in the building. An assistant medical officer and his wife reside here.

G

An unfinished building, which will be three storeys in height, is designed for chronic cases and epileptics.

In addition to these annexes there are no less than five cottages for 25, 20, 33, 16, and 26 patients respectively.

At Norristown, near Philadelphia, segregation has also been carried out to a considerable extent.

Again, at the Concord Asylum, New Hampshire, a very attractive cottage has been erected, at the suggestion of Dr. Bancroft, for a select number of female patients, whose mental condition allows of their living in an ordinary residence. The provision here is on a small scale, and does not profess to provide for the chronic insane.

At Brattleboro, also, in the State of Vermont, Dr. Draper has what is called the "Summer Retreat," built like a Swiss châlet, and accommodating 20 patients. This house cost £145 per bed. At this asylum there is a separate ward for criminal lunatics.

I will next describe the course pursued in another State, Wisconsin, with a view of providing accommodation for chronic lunatics at a moderate cost. There are already two State asylums (Mendota and Winnebago), and what may be called a semi-State asylum (County Milwaukee), for both State and county contribute to its support.* They receive both acute and chronic cases, but they are much crowded, and there is constant pressure upon the authorities for more room. When we consider the vast sums which have been spent upon the construction of the proverbial palatial asylums, amounting in some States to

---

* "Whenever the total number of insane persons in this State shall exceed the total number of such persons who can be conveniently and properly cared for in the State Institutions for the Insane already existing under the laws of this State, the Board of Supervisors of any county in this State, upon the conditions hereinafter named, may purchase or otherwise provide a proper site, within said county, for the erection of a county asylum for the care of the insane and inebriate persons, said site to contain not less than 40 acres; and when said site shall have been approved by the Governor, such Board of Supervisors may proceed, as hereinafter provided, to erect thereon suitable buildings for the proper care of the number of insane and inebriate persons, not less than 30 nor more than 50 per centum greater than the entire number of insane persons then belonging to such county, as such Board of Supervisors may determine." (Law of 1881).

£600 per bed, and this without any extraordinary curative results, we cannot be surprised that guardians of the poor and boards of charities should make a desperate effort to escape such expenditure, and should set themselves to work to provide humbler domiciles for at least the more harmless and chronic classes of patients—those who are, for the most part, regarded as being incurable. Although the cost per bed in the Wisconsin State Asylums would be less than half this amount, the pecuniary element is a strong reason for the course pursued. Thanks to the Wisconsin State Board of Charities, I had every facility afforded me for examining these Institutions; and from this body collectively, as well as from its Chairman, Mr. Elmore, and its Secretary, Professor Wright, individually, I received the most kind and considerate attention while engaged in inquiring into the operation of the system which they have adopted. This may be described in brief as County care under State supervision.

The power to decide whether the counties care properly or not for their insane, upon which depends the all-important matter of an appropriation from the State, rests with the State Board of Charities, and was conferred by the law of 1881, chap. 233, which runs thus:—

"Whenever it shall appear to the State Board of Charities and Reform that insufficient provision has been made for the care and support of the insane in the State Hospitals and County Asylums previously established, &c., the said Board may file with the Secretary of State a list of counties in which no County Asylum exists, and which counties, in the opinion of the Board, possess accommodation for the proper care of the chronic insane ; and thereafter each of the said counties so-named, which shall care for its chronic insane under such rules as the said Board shall prescribe, &c., shall receive 1 dollar 50 cents (6s. 3d.) per week for each person so cared for."

It is obviously important that the State should exercise this supervision over counties in regard to their asylums, but in no other State having them, is this check provided except in New York, and there the Board, although it inspects, does not possess the State-appropriation at command to support its action.

Thus, the counties which take care of their insane under

the authority of the Board, receive 6s. 3d. a week for each person so cared for.* I may state that, during the last year, £9,845 was paid by the State of Wisconsin for this object. Existing almshouses are adapted to the purpose by the Board, or small asylums are built in their neighbourhood. In two counties, however, the asylums are entirely separate, as there are no almshouses, all relief being out-door relief. In the Consolidated Monthly Report of chronic insane under County care in this State, under the provisions of the above-mentioned Act, it is stated that there are 11 counties with asylums varying in capacity from 40 to 100 beds, and containing in all 732 patients. It may be added that 466, or 63 per cent., of these patients were employed, and that during the month four had been discharged recovered or improved out of this chronic class. No patient had been in continuous, but four in temporary, restraint.

I found these small county asylums, on the whole, in a satisfactory condition. The superintendents or masters of the house are laymen of a respectable farmer class, and a medical man in the neighbourhood is engaged to visit at fixed periods, and oftener if necessary. There is a considerable amount of land attached to these houses; and on visiting one of them (the Dane County Asylum at Verona), where there were 97 patients, I found 16, with an attendant, engaged in husking corn. Ordinarily, a larger number work on the land. One patient had, previous to admission, been confined in a small pen in an almshouse, while others had been rescued from neglect or cruel treatment. The patients chop up a great deal of wood in the winter, and slight rewards are given to workers, by way of encouragement. As I

* Last year it cost the counties 11d. more a week per head than they received from the State, without allowing for the investment in buildings and land. It is stated that the salaries of attendants are much the same in County and in State Asylums. The total cost of chronic insane per head per week in the latter is exactly 3·89½ dollars, while in the County Asylums it is only 1·72.

approached this house I observed that the doors and several of the windows were wide open, and that no bars of any kind were to be seen. It has had open doors from the first. In the dietary of that day the patients had coffee for breakfast, with eggs, pork, and potatoes; while for dinner they had beef, potatoes, and parsnips. For supper they had tea or milk, with mush (corn meal), and bread and syrup; some had pie. This institution cost about £6,800, or £68 a bed. In regard to restraint, I found, on referring to the record, that three patients had been restrained in the course of the year. Seclusion had been rarely resorted to. The master, Mr. Myers, evidently felt a warm interest in the patients, and took great pains to induce them to employ themselves. He has a salary of £200 a year. The visiting medical officer has £40.

In another of these asylums (the Dodge County Asylum at Juneau), built at a cost of £60 a bed, and having a capacity for 90 patients, I found a considerable number employed in the potato field, in digging, and in husking corn. Some of the latter were formerly immured in the cells of a wretched "crazy-house," long used in connection with the almshouse. Maize, potatoes, barley, oats, hay, tomatoes, peas, beet, beans, turnips, parsnips, cabbages, and celery, are raised on the farm. Some patients take the entire charge of the cowhouse, and two are employed to milk the cows. Twenty women were employed in the institution, and four in the adjoining poorhouse. The mistress of the house, who was formerly at the old poor-house, gave a graphic description of the condition of the patients who had been in the crazy-houses hard by, which I visited. They are interesting in their present empty state, as relics of the past. Prior to 1871, when the State Board of Charities was organised, the patients used to lie on the straw, either naked or in "slips," generally without any underclothing. The food used to be given to them on a tin plate thrust through a hole in the door. No knives or forks were allowed, and a patient would often throw the food on the straw, and eat it like an animal. The straw was removed from these pens with a pitchfork. When the present treatment was introduced, the patients had to be taught cleanly habits like children, and made to dress, come to table, and go to bed. I saw an epileptic woman, a German, who had been immured in one of these pens. When they dressed her in blue calico she was mightily pleased, and exclaimed, "Schön!" There are ten men and ten women in this asylum who formerly were in the crazy-houses. One day the mistress took an old woman who had been immured there to revisit them. "She was that uneasy and wild, and said, 'Are you going to put me into that crazy-house again? Why don't you burn it up?'" The cost per week, including clothing, is 8s. 9d. for each patient. I was curious to examine the record of restraint and seclusion in this somewhat out-of-the-way institution, and I thought it worthy of transcription:—

June 15th, 1883.—A. B., shut up in room 8 hours for quarrelling. Result, good.

July 5th.—Ditto, 9 hours for quarrelling. Result, good.

July 29th.—Ditto, 16 hours for disobeying and using indecent language. Result, promised to behave in future.

Nov. 8th.—J. L., shut up in room 9 hours, for raising a chair on an attendant. Result, promised good behaviour in future.

Dec. 3rd.—C. D., shut up in his room for 10 hours, for assault on attendant. Result, don't seem to have much, if any, effect.

Sept. 8th, 1884.—C. D., shut up in his room half a day, for striking an inmate. Result, don't seem to mind restraint much.

No women had been in restraint or seclusion.

At another similar institution (the Rock County Asylum, at Johnstown), where there were 75 patients, the master and his wife appeared to be thoroughly interested in their work. Active and successful efforts were made to employ the patients. The first three patients whom I saw were busy in the yard with a cart and potatoes just brought from the field. One patient who had been admitted from a State Asylum, in regard to whom the master was warned that he must only allow him the use of a tin plate at meals, was put to work the day after admission, and the result has been very satisfactory. One patient was out with the team six miles off. I saw sixteen patients working in a potato field without an attendant, a competent patient acting as overseer. They work five hours a day, and I was glad to hear the master observe that it was not wise to let them work until they are sick of it. On examining the record of restraint for the year ending October, 1884, I did not find more than one case of restraint, viz., by mittens for half a day in June, for violence and striking an attendant. Seclusion had been employed five times for four different cases. Wristlets, mittens, and the camisole had been used during the previous year, but to a very slight extent. Two crib beds were in use, one for an idiot, and another for a restless elderly man, constantly getting out of bed. I should be glad to think that they are never used less considerately in any of the State Asylums on the American Continent. It is stated, and I have no reason to doubt the statement, that there is no resort to "chemical restraint" in these County Asylums. A physician, Dr. Rockwell, resides half a mile off, and visits the house nearly every day. The estimated annual value of the labour done by the 75 inmates is upwards of £200. It was estimated that the labour of 16 of the patients was equal to that of the same number of sane persons; that of 20 equal to half the number of ordinary work people, 15 below this mark, and there were 24 who could not work at all.

Many fear that the system thus pursued in Wisconsin for providing for chronic cases will end in grief, as it has done before, and that these institutions in the course of time will become as great a scandal as the old almshouses.

However excellent the present Board of Charities may be (and, in my judgment, is), it is urged that its constitution will change ere long, and that inferior men will very probably be appointed. In reply to these objections it is said that the management of the State Asylums themselves may fluctuate according to the composition of their Boards, and that, in regard to mechanical restraint, its use is at least as great in these institutions as in the county houses. Further, it is alleged that the County Boards of Supervisors, which are assumed to be composed of an inferior class of men, are by no means insusceptible to influence in the right direction from the State Board of Charities, which meets with them and explains the objects in view, and the means by which it considers it necessary to obtain them. Political motives, it is admitted, are a source of weakness and danger.

It is very clear that the success of the system requires constant care in the selection of cases, so as not to place in these small county asylums, where there is no resident doctor, acute and curable cases requiring constant medical care. I believe that, among the chronic insane, there is as a matter of fact very little selection of cases, probably too little. The superintendents of the State hospitals furnish the lists of chronic insane which are to be returned to the counties. They will, of course, retain the best cases, and sometimes send unsuitable ones to the county asylums. Occasionally the counties return a homicidal or filthy patient. Otherwise they have so far taken all the chronic cases from their own counties, who usually are the worst treated, and therefore, probably, the worst behaved, insane inmates of almshouses.

The appointment of similar able and well-intentioned men on the Board of Charities, as well as of thoroughly reliable masters, is also essential to success. Otherwise there will inevitably be a return of the evils from which the insane in the old almshouses have escaped.

It is a noteworthy fact that at the present moment, in

England, the Lunacy Commissioners are encouraging the increased use of workhouses for the chronic insane by the recent action they have taken in regard to sending a number of this class to workhouses from county asylums.

Nothing has been done in America in imitation of Gheel, and there does not appear to be any tendency in that direction.

IV. *Relative Merits of American and English Asylums.*

I wish that I could now convey in a few words, by way of summary, a just idea of the respective merits and demerits of American and English asylums, but this is not altogether an easy, and is certainly rather an invidious, task.

Instead of directly giving the palm to either (though what may be regarded as insular prejudice would scarcely allow of my being discontented with our own), I would say that I believe each has something to learn from the other.

I think, first, that English asylum-superintendents will, in passing through the wards of American asylums, pick up not a few hints in regard to practical details, which they would find very useful indeed. The Americans are so ingenious and inventive a people that it would have been strange had their asylums not borne some evidence of this; and there are many little, but still important, matters in which this inventive faculty is applied to the good working of the institution.

Again, I think the Americans have been wiser than ourselves in avoiding the construction of so many very large asylums.

A third advantage on their side, and partly due to the above fact, is the greater proportion of medical men in most asylums in the States than obtains in England, and consequently the possibility, to say the least, of more individual interest in the patients and their treatment. The proportion varies considerably in different States, but on taking the average of a number of American asylums I found it to be 1 in 150; while on making a similar calculation for those in England I found it 1 in 300. At the New York City Asylum for about 1,500 male patients, there were 15 assistants, several of these being clinical clerks.

Further, I like the more frequent practice of having married assistant medical officers. It is thought to introduce or retain a better class of men, and to give greater confidence to the friends of patients, especially in the absence of the superintendent. As a rule, assistants are better paid than with us. Some receive £300 and £380 a year. Attendants are also very liberally paid; male supervisors receiving from £70 to £100 a year, and ordinary attendants ranging from £50 to £60; female attendants receive £30 to £50; matrons and stewards are handsomely paid. The superintendents themselves are, strange to say, rather under-paid, seeing that they have no pension. I should set the absence of pension against the allowance of full rations granted to the American superintendents, and the liberality of their Committees in paying for their travelling expenses, in many instances, when making even long journeys abroad. Superintendents are often allowed the free use of carriages, horses and servants, so that if they had pensions they would be much better paid than British superintendents.

I may mention here, in passing, that a good plan is adopted among the superintendents of the New England Asylums, of keeping a "black list" of attendants. Thus Dr. Pliny Earle recently warned the Superintendent of the Middletown Asylum (Dr. Shew) in the words, "Beware of Brown." Another superintendent wrote opposite the name of Smith, a discharged attendant, "very bad fellow," and another wrote, "discharged Jones for rough usage of an excited patient;" while in a fourth case the attendant was designated "intemperate."

I am not sure whether or not I should set down to the credit of the State asylums in America, that there are usually some paying patients mixed with the others who are supported by the State or County. It is strongly urged that this helps to raise the tone of the institution, and that the non-paying patients are the gainers by this admixture of classes. It may be so. I have my doubts, however; and rather incline to the restriction of State asylums in America, and County asylums in England, to the pauper class. Certainly, had the American Hospitals for the Insane been intended for paupers only, there would never have been such buildings as the Danvers and the New Worcester Asylums, with their costly furniture and their correspondingly expensive internal economy. Still, there remain serious evils springing from the union of the two classes, and I think I am warranted in saying that Dr. Earle would prefer separate provision for them.

The idea that the State should provide accommodation for the insane of all social classes has long prevailed in America. Dr. Kirkbride maintained very strongly that "there is no justification for a State providing accommodation for one portion of its insane, and leaving the rest uncared for." English opinion, on the other hand, has travelled in just the opposite direction, and the State has not seen its way to provide asylums for private patients. Whether we are coming to this, and are about to Americanize our institutions, remains to be seen. There may have been a tendency in the United States to allow a disproportionate number of paying patients to occupy the State asylums. Until recently many patients of the pauper class were crowded out of these institutions, and suffered from the gross neglect which befell them in almshouses. Now that this evil is to a considerable extent rectified, the friends of the higher-class patients require more provision for their relatives, and the consequence is that private asylums are on the increase. In a number of States, *e.g.*, in Ohio, Indiana, Illinois, and Minnesota, there are no pay patients. The institutions are free to all residents of the State who are insane, regardless of their pecuniary condition. The wealthy are supposed to have paid their taxes for the maintenance of their asylums, and to be entitled to their benefits without further charge, just as they are to the benefits of the public school system.

It has been said that the classification of the insane in asylums is more perfect in America than in England, but I cannot say that I was struck with this difference myself, although the arrangements for dividing the corridors at will by folding doors are especially good in some asylums. Dr. Kirkbride had eight separate wards in his asylum, but I do not think that these divisions resulted in much, if any, more practical differentiation of cases than obtains in our asylums; and, indeed, the rule which he laid down as the basis of classification—the bringing into the same ward patients who would consort well together, and the separation of those who would not—had too much common-sense and simplicity in it to allow of great complexity of arrangement.

I think I ought to mention the development of the system of the segregation of patients, to which I have referred as at the present time a feature in American practice; for although detached buildings for different classes have been long advocated and used in British asylums,* and were at one period discountenanced by alienists in the United States, the separation of cases differing in their mental character, and therefore requirements, has now been carried to such an extent, and as some hold to such an extreme, that it really forms a marked feature of recent and present movements among the promoters and organizers of asylum provision in the States. I think, therefore, that this strong tendency to

---

* It was originally intended to build the new Worcester Asylum (Mass.) in detached buildings, to the same extent as at Whittingham (Lancashire), but the intention was abandoned.

segregation, which has manifested itself, must be regarded as possessing greater momentum in America at the present moment than even in this country, and that we shall be able to learn something, or be confirmed in what we have already learned, by the plans now being carried out in some of the States.

I am, on the whole, disposed to reckon among the advantageous courses pursued by the Americans, the appointment of lady-physicians in some of their asylums —a practice which is certainly growing. I regard it as an experiment, and I think we ought to be grateful to our friends across the water for making it. I would go further, and say that if the lady-doctors of the future should be equal in ability and high moral character to those who have hitherto held office, and, if their position is so clearly marked out as to prevent all clashing with other members of the medical staff, they will prove a decided blessing to the female patients in asylums, and a real help to the medical superintendents.

Before leaving the congenial task of pointing out the merits of the institutions of our Transatlantic brethren, I must not omit to mention the favourable impression I received as to the diet allowed to the patients. I consider it more liberal and as having more variety than in our own asylums.

It is probable that in the main the Americans not only feed, but house and warm, their patients better than we do, while individual comforts, likes, and dislikes, receive rather more recognition and attention in their hospitals than they do in ours; this is due, no doubt, partially to national differences, and, of course, the same standard will not apply to both countries.

As to the comparative demerits of the American asylums, I would repeat that in some, especially the old American institutions, there is not so much employment of the patients as there might be, and as, for the most part, is carried out in those of our own country.

On another point I think that British asylums show to better advantage than those on the other side the Atlantic. I was struck with the generally bare, unfurnished condition of the galleries occupied by the excited patients in the latter. I never saw, for instance, such comfortable-looking quarters for this class as those provided in the charitable Hospital of Bethlem, or, to take a pauper asylum, in that at Prestwich; in both instances old asylums are made to look homelike and comfortable by the pains taken to furnish them in the way English people like to have their own houses furnished and decorated. There are, no doubt, some rooms in some American asylums which would not only equal, but surpass, those in our asylums, whether Registered Hospitals or for the County, and this must account for two of my American friends having been struck in exactly the opposite way, for they specified the bareness of English asylums as one chief point in which they are inferior to their own.

Then, of course, in regard to mechanical restraint, it will be inferred from what I have already said, that I regard the lesser amount of restraint in British asylums as preferable to the greater amount of restraint in those of the United States. Still, notwithstanding this criticism, there is a growing tendency on the part of superintendents of American asylums to trust the patients with more liberty, and to remove unnecessary signs of forcible detention, and unsightly means taken to protect glass, &c., from injury. Thus, in some instances, they have removed from the corridors iron frameworks which were formerly used to protect the windows, a few feet from (and within) which they were placed. The panes themselves have been made larger and the bars less frequent, while small glazed apertures high up in single rooms have been lowered and enlarged. I am now speaking of the main buildings. In the annexes and cottages, anything partaking of the character of a prison is for the most part altogether absent. All this indicates a healthy feeling and breadth. It shows, along with the breaking-up of one monotonous building into several, that whatever faults there may have been in past times in regard to the sameness of asylum wards, there is at the present time a vigorous effort to adapt the construction of the buildings occupied by the insane to the very various mental conditions which they present.

In conclusion I would say that the outlook with regard to the future of the insane in the United States is very encouraging.

To say that the asylum-physicians of America have not utilized, to the extent they ought to have done, the materials at their command; that their annual reports are defective in scientific results; that they have made no great important discoveries in the treatment of insanity; is, alas, only to bring a charge against them which is frequently brought against the superintendents of asylums in our own country. In both countries those who are thus charged will sorrowfully admit that there is only too much truth in the criticism; but they will, while pleading guilty to their shortcomings, ask for an indulgent judgment on the ground of the mass of routine work which falls to their lot. With regard to American asylums, there is, at Dr. Gray's Asylum, Utica, a special pathologist, Dr. Deecke, whose micro-photographic sections of the whole area of the brain and cord are well known. There is also a pathologist at the Washington Asylum. At the Middletown Asylum, Connecticut, under Dr. Shew's care, much attention is paid to microscopic work; while I may mention that at the Norristown Asylum, near Philadelphia, a lady-physician performs pathological duties in the institution. Outside asylums, cerebral morbid anatomy has been pursued by neurologists and alienists, but I must restrict myself in these remarks to hospitals for the insane.

The American authorities have had enormous difficulties to contend with from the fact of society in America being in a state of continual fusion, including the mixture of races consequent upon immigration. When we are disposed to condemn the long-continued lamentable neglect of lunatics in almshouses, and the slow progress made in improving the condition of some of even the large asylums, especially the institutions in the Western States, and their absence or insufficiency in the Southern States, we ought to remember that a very large mass of insanity has been thrown upon the American townships, Counties, and States, very much against their will. We can well understand how this must have been the case when we learn that, between 1820 and 1850, 2,250,000 emigrants landed in the United States, making one-tenth of the population foreign. The number of insane in 1850 was 15,610, and of these, 2,049, or very nearly one-seventh, were foreigners. During the succeeding ten years, immigration increased so much that there were, in 1860, 4,136,000 foreigners, or about one-eighth of the population, while the insane foreign element amounted to 5,768, or one-quarter of the number of insane in the States (24,000). Again, travelling over the ten years between 1860 and 1870, we find that at the latter date the foreigners amounted to 5,567,000, or one-seventh of the total population, while of the insane (37,432), no less than one-third were returned as foreigners. Lastly, taking the period from 1870 to 1880, during which praiseworthy efforts have been made to alleviate the condition of the insane in almshouses, we find that at the latter date there remained about one-seventh of the population foreigners. As we have already stated, the census of 1880 showed the number of insane to be 91,959; of these, no fewer than 26,346, or between one-fourth and one-third, were not American born. In other words, 13·33 per cent. of the general population—that is to say, the imported element—produced 28·75 per cent. of so-called American lunacy. Stated in another form, if the native

Americans* alone are considered, there is one insane person to every 662 of the population; while the proportion among the foreigners alone is as high as one in 250. To show that the liability to insanity among foreigners in the States is greater than that of the native population, and that therefore the Americans are disproportionately weighted with this terrible burden, it may be added that if the average proportion of insanity among native whites were the rule, the number of insane in the United States in 1880 would have been 81,158 instead of 91,959.†

The force of the present contention would be much strengthened were we to include the hereditary influence exerted by the foreign class—one which does not appear in any statistical tables, for the insanity which occurs among the children of mixed marriages will be merged in that of the native population. It is hardly necessary to say that the Americans, while they are becoming alive to the heavy burdens inflicted upon them by immigration, in the form of insanity, pauperism, and crime, are not blinded by these facts to the great balance of gain which is theirs. They feel that they must arouse themselves to keep out the scum and dregs of European populations, and that the check must be made in the first instance at the port of departure; also, that unsuitable immigrants should be forbidden to land, by the control of quarantine in charge of medical experts. A bureau of health and immigration should, it is proposed, have agents abroad attached to consular offices, and agents in the ports of the States to carry out the powers granted them by Congress. Dr. Foster Pratt, who in his excellent paper makes these and other proposals, writes in no bitter spirit in regard to

---

* Including the coloured population, among whom the ratio of insane is as 1 to 1,097. As is well known, idiocy rather than insanity prevails among the negroes.

† These figures are given by Dr. Foster Pratt in a paper read before the American Public Health Association in 1883, and are based on the last Census. Of the 91,959 insane, no less than 88,665 are paupers, of whom 22,961 are foreign-born.

immigration, and would by no means wish to tax the immigrants to pay the expense involved.*

I ought to add that Mr. Wines, while admitting the correctness of Dr. Pratt's figures, does not altogether agree with his deductions. He writes " insanity is a disease of rare occurrence prior to puberty. But the native population includes a very much larger number of children, proportionally, than the foreign population does. If we compare the insane natives and foreigners with the population above the age of 14, we shall secure a truer result :—

Native whites	... 59,645 :	22,002,165 = 1:369
Foreign ...	... 26,158 :	6,196,125 = 1:237
Coloured ...	... 6,156 :	3,918,595 = 1:637

" According to these figures, the tendency to insanity on the part of the foreign population, instead of being, as Dr. Pratt makes it, 'three of foreign to one of all white natives,' is about one and a half of foreign to all white natives."

I will only add to the observations now made, that if I have not hesitated to speak freely of any shortcomings, as they seem to me, in the arrangements or administration of the American asylums, I entertain the hope that their virtues rather than their failings will attract our notice, and that we shall strive to emulate their example in everything in which they have succeeded in adding to the convenient appliances of asylums, or in increasing the comfort or happiness of the patients confined therein. Fresh from the unstinted hospitality and attention received from American physicians and others, I would express the desire that they may always receive a hearty welcome from the medical superintendents of English asylums, when they visit our shores.

* At the present time all aliens arriving at any port in the United States must pay a tax of 50 cents, which goes towards the expenses incurred on the arrival of distressed immigrants, but does not help the interior States to care for them when they become insane.

CHAPTER IV.

Principal Asylums Visited.

Having in the preceding chapter given a general sketch of the prevalent views and practices in regard to the care of the insane in the United States, I proceed to note, very briefly, at the risk of occasional repetition, the salient features of the principal institutions visited.

The first institution I visited was the *New Hampshire Asylum for the Insane* at Concord. It is a "Corporate" institution. The government is vested in a Board of Trustees, twelve in number, who are appointed by the Governor of the State and the Council. A Board of Visitors inspects the hospital at stated periods, and reports biennially to the Senate and House of Representatives.

Dr. Charles P. Bancroft is the Medical Superintendent. His much-esteemed father, Dr. Bancroft, who held the office till recently, now resides near the asylum, and continues to take an interest in it in various ways. I began my visitation of the asylums of America with the same kind and hospitable welcome which was extended to me to the last day of my stay in that country.

The number of acres at the Concord institution is 130, which include an excellent farm of great utility to the patients. These are 320 in number—130 males and 190 females. Socially they are a mixed class. The State contributes 16s. a week for free patients. The weekly charge varies from 12s. 6d. to £4. About one-sixth do not pay anything. Mentally, the patients consist of acute and chronic cases. As many as 284 are incurable. Six men and one woman labour under general paralysis of the insane.

The proportion of attendants to patients is as 1 to 9; in the refractory wards, 1 to 7. Their salaries vary from £4 to £6 a month for male attendants, and £2 12s. to £4 4s. for female attendants.

With regard to restraint and seclusion, I found that on the male side of the house three patients were in restraint, and thirteen on the female side. Dr. Bancroft said that some of the restraint could be easily dispensed with if more attendants were allowed. One man and two women were in seclusion.

Of 4,614 patients admitted during the 41 years the establishment has been in operation, 1,711 have been discharged recovered, or 37 per cent. During the same period 714 patients died, but, as the average number resident is not stated in the statistical tables, I am unable to give the mean annual mortality per cent.

Beyond the general comfort and satisfactory condition of this institution, the most noteworthy circumstance at the present time is the new building for quiet and convalescent female cases, with a capacity for 25 to 35 patients, according as they have single rooms or not. This house is very tastefully built, and is made to look as much like a private dwelling-house as possible. The cost of this building has amounted to something more than £2,000. It is fully intended to erect a similar house for male patients when the necessary funds are forthcoming.

I was glad to find that the friends of patients who are ill are encouraged to come into the house and stay with them; for which purpose certain rooms are arranged so as to allow of great privacy, and present a very comfortable appearance. Nearly all the bedrooms are single, the patients much preferring them to associated dormitories.

At the time of the Census (1880), the following were

the numbers of insane and idiots, and their location in this State:—

	Insane.	Idiots.	Total.
New Hampshire Asylum	288	5	293
Almshouses	261	155	416
Jails	0	0	0
At home	507	543	1050
	1056	703	1759

The population of New Hampshire in 1880 was 346,991.

Dr. Bancroft informed me that in New Hampshire there are not a few almshouses, or, as they are otherwise called, county-houses or farms, in which there are lunatics. The County Commissioners, who are like our Poor Law Guardians in England, have authority to send pauper lunatics to these houses, and no medical certificates are required. Their condition is very bad indeed. Seven years ago Dr. Bancroft drew attention to the necessity of the State providing for all its pauper insane; and in 1883 a Commission, appointed to inquire into the subject, published a Report, from which I will make a few extracts, as they present a correct picture of the present as well as the past condition of things, and as this lamentable state of lunatics in almshouses or county-farms is the crying evil of New Hampshire, and of any other States in which patients requiring the treatment and care of an asylum are detained by the County Commissioners in almshouses.

From this Report it appears that in the county-farms above mentioned there were on February 1st, 1883, 488 paupers of unsound mind, of whom 134 were either feeble-minded, idiotic from birth, or demented, while 354 were considered insane in the popular acceptation of the word, nine of whom were regarded as curable. It will be seen that, since 1880, there is an increase of insane in these houses.

No special treatment is attempted at the almshouses in seven counties. "In truth," says the Report, "special treatment of mental disease, as such, is nowhere attempted."

On every county-farm were buildings, or parts of buildings, devoted to the insane; seven of these being of wood. Some were fairly adapted to their purpose, while others were only the upper storey of an open wood shed, or a half-underground cellar. Some patients were "found in solitary confinement, night and day, summer and winter, in what it is no exaggeration to call cages, to save the expense of suitable attendance."

In these almshouses were 121 strong rooms for lunatics requiring confinement all or a part of the day. The number so confined was 81. In three of the almshouses, insane persons were locked up at night by themselves, without the presence of any attendant in the buildings devoted to their occupancy.

The cost per week to the county for keeping an insane pauper averaged about 5s. 4d. each, which, the Report observes, "will hardly excite admiration in the minds of well-informed friends of the insane." It is added, however, that this does not include all the farm products devoted to the inmates of county-farms.

The Report points out that the entire lack of system in the care of the insane paupers of New Hampshire arises from defective legislation, and instances the fact that they are allowed to mingle with the sane, and sometimes even with criminals, in the jails and houses of correction. It is pointed out as a singular anomaly that, while the State provides that insane persons sent to the New Hampshire Asylum shall be certified, before admission, by two physicians, and the institution shall be carefully inspected by a Board of Visitors, and by a Board of Trustees, and attended by skilled physicians, no provision is made to guard against the improper admission or detention in almshouses of persons alleged to be insane, and no directions are given as to their treatment by skilled physicians. Neither are reports of their condition, nor such statistical information as is universally required of the managers of institutions for the insane, demanded.

It is an unfortunate circumstance that the County Commissioners are selected from political considerations, at warmly-contested popular elections.

The Report recommends the establishment of a Board of Commissioners in Lunacy, consisting of three persons conversant with the treatment of the insane, to be appointed by the Governor with the advice of the Council; these Commissioners being empowered *inter alia* to send all such paupers in the county-almshouses as are curable to an asylum for the insane, to be there supported at the expense of the State; no insane person not a pauper being allowed to reside in a county-almshouse.

It is recommended further that a State-asylum for insane county paupers deemed incurable, be erected and maintained at the expense of the State, and placed under the direction of the Lunacy Commissioners in charge of a superintendent appointed by them. Only those lunatics who can, in the judgment of the Commissioners, be as well cared for at the almshouses as in asylums would be allowed to remain there.

It is greatly to be hoped that steps will be promptly taken to provide proper accommodation for the insane and idiotic in New Hampshire, for, as will be seen, out of nearly 1,800 of these classes there is asylum-room for only about 300.

The next asylum visited was the *Vermont Hospital for the Insane* at Brattleboro. The salient characteristic of this well-managed institution, under Dr. Draper's charge, is the "New Park" for the use of male patients, 32 acres in extent, laid out about three years ago. This beautiful wood has been more recently brought into use, and has afforded employment for the patients. Then there is the "Summer Retreat," with its 20 acres of ground, for 20 female patients. It was opened a couple of years since, and it is occupied about four months in the year. The doors are unlocked. There is also an excellent gymnasium, 75ft. by 40ft., where bowls, quoits, billiards, and dancing are enjoyed, as well as gymnastics. This is an excellent place for indoor recreation in winter. Lastly, I may mention the promenades for women, which, sheltered from the weather, have been built for their use within the last two years. There is a separate ward for criminal lunatics, of whom there were 19; ten of these being convicts who have become insane. The dimensions of the single rooms are eight by ten feet, the height being also ten, thus giving 800 cubic feet. Solid wooden bedsteads are secured to the floor. The grated shutter or screen, in common use in American asylums, protects the windows. Attached to this ward is an airing-court, in which most of the patients of this class were walking about. Four were in restraint, chiefly by means of the leathern muff, not altogether without reason I must admit, although at Broadmoor they would not have been mechanically restrained. Dr. Draper would probably say that, as both the superintendent and the assistant medical officer of the latter institution have recently been seriously injured by the assaults of criminal lunatics, he does not think Conollyism altogether so successful as to be tempted to adopt it. Dr. Draper, however, is by no means prejudiced against non-restraint. He visited our country in 1881, and determined, on his return, to disuse restraint entirely, but he hesitates to do this in homicidal cases. He does not doubt it could be disused if the number of attendants were increased, but, to this, objection would be made on the ground of expense, unless he were decidedly of opinion that it was essential.

The government of the hospital is vested in a Board of Trustees, four in number, who report to the Governor of the State.

A Board of Visitors, six in number, periodically inspect the institution. The asylum is also inspected by the

State Board of Supervisors, three in number, one of whom is required to visit monthly. They report biennially to the Governor.

The number of acres amounts to 600; 200 being under cultivation, the rest pasture and woodland.

The number of patients is 436; males 236, females 200. This excess of males is striking. Dr. Draper is unable to account for the disparity (which has been marked for 15 years) unless it be that the prevailing type of insanity in the females is quieter than in the male sex, and so does not require removal from home to the same extent. In the general population of the State, females are in excess of males. The classification mainly depends upon the mental condition of the patient. There are nine wards on each side of the house. Socially the patients are of a mixed class, mainly agricultural. The weekly charge is £1, and the cost 16s. 6d., including "permanent improvements;" without them, 15s. An additional allowance of 2s. 2d. is granted for State patients.

The proportion of attendants is 1 to 11, and their salaries are as follows:—Males, £4 10s. to £5 per month; females, £2 6s. to £3 10s.

As to restraint and seclusion, there were on the day of my visit, the four men I have mentioned and two women restrained on account of suicidal and homicidal tendencies. This, Dr. Draper stated, was a fair average. There were in seclusion 2 men and 1 woman.

With regard to recoveries, Dr. Draper's experience agrees with that of those who do not find the proportion of recoveries, stated, 25 years ago to have been about 42 per cent., is borne out by more recent statistics. He thinks that mental disorder tends to assume a more permanent form. Of 5,892 admissions since the opening of the asylum in 1836, 2,354 have been discharged recovered, or 39 per cent.

It is impossible to calculate the mortality on the mean number resident, from the statistical tables in the Report.

The number of deaths has been 1,427 since 1836. In 1883 the ratio of deaths to the average number resident was about 7 per cent. Some die every year from typhomaniacal exhaustion. General paralysis has recently increased the mortality. However, out of 186 admissions in 1883 only 5 were paralytics.

In regard to occupation, I found that 33 per cent. of the men were employed out of doors, and the rest were generally able to go out in the grounds. About the same proportion of women were employed indoors. Dr. Draper strongly insists on the importance of exercise in the open air in mania without exhaustion.

I may say, in conclusion, that it is difficult to convey an adequate idea of the beauty of the grounds and surrounding hills of the Vermont Asylum. English medical superintendents should supply the defects of my description by a personal visit. If they do so they will not be disappointed in either the scenery without, or the indications of effective management within.

The population of Vermont is 332,286, and the returns of insane and idiotic in 1880 were as follows:—

	Insane.	Idiots.	Total.
Brattleboro Asylum	454	2	456
Almshouses	45	88	133
Jails	0	0	0
At home	516	713	1229
	1015	803	1818

Dr. Draper does not think that the lunatic poor in Vermont, who are still in almshouses, suffer. They are chiefly superannuated and dements, and there are very few of them. If a patient were to become much excited, he would be transferred to the Brattleboro Asylum. At the same time Dr. Draper sincerely hopes that the State of Vermont will build an asylum for its pauper lunatics. A clause in the original deed of the Brattleboro Asylum appears to render it incumbent upon the trustees of this asylum to take charge of at least a certain number of poor patients.

Passing from Vermont to Massachusetts, I visited, first, the *State Hospital for the Insane at Northampton.* The name of Dr. Earle, the head of this institution, is almost as well known to English as to American alienists.* He has held office here for 21 years, and was formerly physician at the Bloomingdale Asylum. He and Dr. Butler are the only survivors of the original thirteen who founded the American Association of Superintendents of Hospitals for the Insane, and he is now the President of that Society.†

Among the noteworthy features of this hospital is its financial success. No doubt this might be associated with a far from satisfactory condition of the patients; but when, as in the present instance, the reverse is the case, this aspect of the institution is worth recording. In the latest annual Report of the State Board of Health, Lunacy, and Charity, it is observed that " the hospital is in much better condition than when Dr. Earle took charge in 1864. The farm has been nearly doubled in size, and very largely increased in fertility; well fenced, with its farm buildings greatly enlarged. . . . The new storehouse has cost about £3,000, and all this has been taken from the current surplus and income of the hospital without drawing on the State Treasury. In this respect the Northampton Hospital has set a good example for the other State hospitals to follow, having, by economy of management, accumulated a fund from which it has been able for many years to make all the improvements necessary, without asking for an appropriation" (lxxxix.).

That economy has not been carried out at the sacrifice

---

* Since the above was written Dr. Earle has resigned, and the senior assistant medical officer, Dr. Nims, who has recently visited England, succeeds him.

† Office resigned at the recent Annual Meeting of the Association, at Saratoga.

of the dietary of the patients, appears from the following *menu*, which is not, I believe, surpassed by that of any other State Hospital:—

BREAKFAST.

Monday and Thursday.—Tea, coffee, broiled beef-steak, potatoes, warm rolls, bread and butter.

Tuesday and Friday.—Same, with the exception of fried tripe instead of beef-steak. (In winter, sausages, in spring, fried ham and eggs.)

Wednesday.—Broiled mackerel is substituted for tripe.

Saturday.—Either fried fish-balls or liver, meat hash, and hot corn-cake.

Sunday.—Cold corned beef, potatoes, warm rolls, bread, butter, and fried Indian corn pudding.

DINNER.

Monday.—Roast beef, potatoes and one other vegetable, bread, butter, boiled rice with syrup or sugar.

Tuesday.—Vegetable soup, roast or stewed veal, potatoes and one other vegetable, bread, butter, and baked Indian pudding.

Wednesday.—Either fried or baked fresh fish or boiled mutton, potatoes and one other vegetable, bread, butter, and berry or apple pudding, with sauce.

Thursday.—Vegetable soup, corned beef, potatoes and one other vegetable, bread, butter, and boiled suet pudding with syrup.

Friday.—Either boiled or roast mutton, stewed or roasted veal, potatoes and one other vegetable, bread, butter, and tapioca pudding, or raisin pudding of either rice, bread, or cracker.

Saturday.—Baked beans, corned beef, potatoes and one other vegetable, pickles, bread, butter, and baked bread pudding.

Sunday.—Cold corned beef, potatoes, warm baked beans, pickles, bread, butter, and pies.

SUPPER.

Monday.—Tea, bread, warm corn cake, butter, hard gingerbread, and a relish.

Tuesday.—Tea, white bread, Graham bread, butter, soft gingerbread, and a relish in the warm season, replaced by buckwheat cakes in the cold season.

Wednesday.—Tea, bread, butter, cookies and ginger snaps, and a relish.

Thursday.—Tea, bread, butter, pie, and cheese.

Friday.—Tea, bread, butter, cake, and a relish.

Saturday.—Tea, bread, butter, dough-nuts, and cheese.

Sunday.—Tea, bread, butter, cookies and ginger snaps, and blanc-mange or corn-starch.

For invalids and others who are not able to take the regular fare, beef tea, chicken broth, mutton broth, scalded milk, arrowroot-gruel, oatmeal-gruel, milk-punch, cracked wheat, oatmeal-porridge, dry toast, boiled eggs, &c., are provided.

Secondly, the amount of work performed by the patients has been a point of special attention with Dr. Earle, but as this has been already dwelt upon in the preceding pages, more need not be said about it in this place. It deserves remark, however, that in the table of the number of patients employed, given in the last Annual Report, it appears that all these patients worked as many hours as the employés with whom they were respectively engaged.

Not the least remarkable feature of the indoor management of the patients is the fact that they assemble in the asylum chapel almost every evening during the year, for purposes of instruction, entertainment, and amusement. Dr. Earle is convinced that the labour and expense thus bestowed bring a liberal recompense in the increase of contentment and satisfaction of the patients, and in their greater quietude, self-control, and orderly conduct, not only during these meetings, but at other times and places. Dr. Earle and his staff have taken an active part in the readings and the lectures upon a variety of subjects calculated to interest the patients.

The hospital is governed by a Board of Trustees, five in number, which reports annually to the Governor of the Commonwealth of Massachusetts, and the Council.

In addition to the visits of the Trustees, the hospital is inspected by the State Board of Health, Lunacy, and

Charity through its inspector, Mr. Sanborn, nearly every month.

The first cost of building was £158 per bed, land and furniture included; there being 340 acres.

The number of patients is 460—227 males, 233 females.

Socially, the class of patients is mixed. Dr. Earle endeavours as much as possible to avoid drawing a line of demarcation between the private and public cases. The weekly average number of State patients is 161; town patients, 247; private patients, 57.

Character of patients, mentally; acute and chronic, chiefly the latter. The number of recent cases is, indeed, exceptionally small.

The cost per week is 13s. 6d., or £35 per annum. This amount is calculated by dividing the cost for the year (82,000 dollars) by the average number of patients, 466·76. If the total expenditure for the year (89,523 dollars) be divided by the average number of patients, the result is £38 4s. 8d., or a weekly average of about 15s. 3d. The previous calculation was obtained after deducting the extraordinary expenses from the year's expenditure, and allowing for certain assets.

The weekly average charge per head for all the patients is 14s. 6d. Although a State institution, it has received no gratuitous assistance since 1867, having relied, for its income, upon the farm and the board-bills of patients. For the entire support of State patients, including clothing, breakage, &c., the hospital received from the State 14s. 6d. per week for each patient, from 1870 to 1879.* For one year after, it received 12s. 6d., and since 1880 it has received 18s. 8d., the compensation fixed by statute law. For town patients the same weekly sum is received, but the towns clothe their patients and pay for damages. For private patients, the average payment

---

* It may be stated that in Massachusetts, the counties, as political organizations, have no paupers. There is one county asylum, but its patients are supported by the towns or cities to which they respectively belong.

amounts to somewhat more than £1 a week, clothing and damages being extra charges.

The proportion of attendants is 1 to 12. Their salaries are—Men, £4 4s. to £9 per month; females, £2 12s. to £5.

With regard to restraint and seclusion, Dr. Earle has never committed himself to unreserved approval of the non-restraint system, but the amount of restraint is small. There is one singular case, that of a man who insists upon having his wrists secured, and declares that if they are free he will break everything within his reach. The door of his room is open, but he refuses to leave it. He occupies himself in writing, and drawing diagrams. For several years he refused to wear any clothing; latterly he has worn a long cotton shirt. He objects to having his hair and beard cut, and they have been allowed to grow for 18 years. He is, notwithstanding, clean and neat. Seclusion is resorted to, but not frequently. No covered bed was in use when I visited the asylum, nor has there been for the last two years. One without the cover is used by a bad epileptic.

Coming now to recoveries, I have already referred to the work done by Dr. Earle in regard to the curability of insanity, and need not repeat it here. It is to be regretted that the series of statistical tables prepared by the Board of Health, Lunacy and Charity in 1880 contains no table showing the total number of recoveries for a longer period than the preceding year, so that unless one has a series of reports at hand, it is impossible to ascertain the experience of an asylum for more than one year, which is worthless. Dr. Earle informs me, however, that from the opening of the asylum to September, 1884, the recoveries have been 22·65 per cent. of the admissions, this low percentage being largely due to the fact that, the hospital being originally too large to be filled by patients from the locality, it was for nearly twenty years made the receptacle for the overflow of the hospitals of the eastern section of the State, most of those thus received being incurable. He adds that of the persons directly committed to the hospital prior to September 30, 1870, the recoveries were 29·61 per cent.

The mortality, calculated on the average number of patients for 25 years (1858 to 1884), was 7·72 per cent. For the first 13 years it was 9·05, and for the last 13 years it was only 6·69, a reduction which is attributed to improvements in the hospital and in the details of the treatment pursued.

As to treatment, I am not aware that Dr. Earle has any specially favourite drugs in the treatment of the insane. He was early opposed to the depletion of patients when formerly so much in vogue in America, as already stated (p. 21). Chloral is not administered. Cod-liver oil, bark and iron, and other tonics are freely used. Neither prolonged, shower, nor Turkish, baths are employed.

The number of cases of general paralysis is very small, viz., two, one being a woman. The paucity of cases is attributed by Dr. Earle to the remoteness of Northampton from the large centres of population.

The importance attached by Dr. Earle to the occupation of patients has been already shown by the number employed in the asylum. In the Annual Report, the farm is stated to be one of the important means employed in the hygienic and restorative treatment of the patients, and to afford no inconsiderable source of income—considerably upwards of £2,000 per annum.

As I shall have to mention that a large number of strangers are allowed to visit some of the American asylums, I may add that the practice of admitting the public indiscriminately to the wards of the Northampton Hospital was abolished more than 10 years ago.

After spending several days at Northampton, and attending the meeting of the New England Psychological Society at Boston, referred to at p. 76, I visited the *McLean Asylum* at Somerville, on the outskirts of the city, accompanied by Dr. Baker. It possesses about 100 acres of land.

It might have been supposed that there would not be any very novel feature to point out in an old institution like the McLean Asylum. There are, however, not a few points of interest. When Dr. Bucknill visited it in 1875, he was informed that before long it would be removed to the country. Unfortunately this intention has never been carried out, and the condition of always expecting to move, but never moving, cannot fail to produce a somewhat depressing effect upon the managers and staff. In spite of this, however, the efficient Superintendent, Dr. Cowles, has introduced many improvements, and succeeded in making the corridors and rooms of the present structure as comfortable as possible. Many of the rooms were handsomely furnished, and the patients had the appearance of being well cared-for in every way. I may remark, in passing, that the distinguished Dr. Bell, who has given his name to one form of insanity, was formerly Superintendent of this institution. The asylum is governed by a Board of Trustees of the Massachusetts General Hospital, with which it is associated; and it is inspected by the State Board of Health, Lunacy, and Charity at irregular periods, generally twice a-year.

The number of patients is 170; 95 males and 75 females. They belong mostly to the upper classes of society. In a recent Report it is stated that, of 82 persons admitted, 53 were recent cases, and 29 chronic or incurable.

The cost per week is on an average as high as £3 per patient. There is a considerable annual deficiency, which is made up from a special fund. The highest payment is £15 a week; several patients pay £10. A common payment is 28s.; a few pay 48s. a week.

The proportion of attendants to patients is 1 to 3; 17 persons are on duty at night—including these, the proportion is nearly 1 to 2. Their salaries per month are— males, £4 12s., rising to £6; females, £2 16s. to £4. Those in special wards, and the females in the men's department, get £1 extra a month, or £60 a-year.

Dr. Cowles attaches great importance to the employment of female nurses in the male wards. He believes this to be as easily managed as in a general hospital. He holds that it is unnatural for patients to be placed in asylums under entirely different conditions, in regard to female society, from what they have been accustomed to at home. They degenerate in speech and conduct, and the male attendants are also injuriously affected. The presence of female nurses restrains and softens the insane, and enforces self-control.* All this may be true, but it is obvious that the practical carrying out of the system must add greatly to the anxiety of the Superintendent, and that it should be introduced only by one who is confident that he can carry it out successfully. Nor should it be introduced into all asylums. I would add as a valuable feature of this asylum that it forms a training school for nurses. This is probably rendered easier from its close connection with the general hospital.

Several patients were restrained and in seclusion.

In regard to recoveries, Dr. Cowles, in reporting the percentage in one year as 30·48, observes that this is a larger proportion by five than in any year since 1870, which is partly attributed to the Danvers Asylum being full, and therefore the McLean Asylum receiving more acute cases than before.

Of 6,604 admissions since the opening of the institution in 1818, 944 have died; but no table shows the mortality on the average number resident for this period. Since 1837 the percentage, so calculated, is about 10.

Dr. Cowles makes some discriminating remarks, in his Report, on the use of *massage* in insanity. He observes that, in his experiments during three years with the "rest treatment," he had found it necessary to modify it, some cases having their mental depression increased thereby, while others with great nervous irritability and

---

* Dr. Cowles might take for his text, *Emollit mores, nec sinit esse feros.*

anguish found that enforced rest was intolerable—indeed, positively harmful. He observes: " While rest is useful in some conditions of melancholia, &c., and benefit has been derived from massage, by producing tissue waste and improving nutrition, and from the other means of treatment, 'seclusion' has been almost entirely abandoned, as not useful, and as depriving the patient of what is beneficial." He quotes from a letter addressed to him by Dr. Weir Mitchell, in which he says that in the treatment of "not a small number of cases of melancholia, with bad nutritive break-down, in which I attempted to relieve by rest, &c., I made some successes, but more failures—made, in fact, so many that I gave up at last the effort to treat in this way distinct cases of melancholy. . . . I may use massage or electricity in melancholy, but I do not seclude or rest these cases."

The visit to this excellent institution was somewhat hurried by our having to be present at the American Social Science meeting held at Saratoga on Sept. 12. Here the lunacy laws were discussed, and the opportunity occurred for a friendly interchange of sentiment on the legislation for the care and protection of the insane in Great Britain and the United States. Professor Wayland presided, and in the absence of Mr. F. H. Wines, Mr. Sanborn read a paper by the latter on the laws of lunacy in the different States of the Union, for which the author expressed his indebtedness to Mr. George Harrison, of Philadelphia, whose useful compilation is widely known.* It proved an interesting reunion.

We proceeded from Saratoga to Utica, where we were met by Dr. John P. Gray, whose asylum, the *New York State Hospital for the Insane*, next claims our attention.

* " Legislation on Insanity," &c., by Geo. L. Harrison, LL.D., late President ef the Beard ef Public Charities of Pennsylvania. Philadelphia, 1884. Dr. Charles F. Felsen, of Besten, has also issued a very useful beek, entitled " Abstract ef the Statntes ef the United States, relating te the Custody ef the Insane." Philadelphia, 1884. Dr. Folsen is the Assistant Professor of Mental Diseases, Harvard Medical School.

The characteristic feature of this institution is Dr. Gray himself, who for 30 years has been the Superintendent. His distinctive personality has long made itself felt in the States, and, of course, more particularly in the asylum of which he has been Medical Superintendent for so many years.

The well-known Dr. Brigham was the first superintendent.

One of the noteworthy arrangements of the institution, though not peculiar to it, is the employment of a pathologist, who resides in a separate house, and has every means placed at his command for carrying out his observations and microscopical examinations. I have already mentioned Dr. Deecke as holding this post, and need not again refer to this aspect of the hospital.* It is surprising that out of between 50 and 60 deaths in a year only about a dozen post-mortems should be obtained.

The entrance to the asylum, which was opened in 1843, forcibly recalls the stone columns and portico of Bethlem Hospital. The central building consists of four storeys; the wings, three in number, being of different heights and having one, two, and three storeys. A building is being added for female patients of the excited class, which is well adapted for the purpose. The equable temperature of the asylum is well indicated by a chart, showing a marked difference between the outer and inner ranges of temperature. The fan system of ventilation, to which the American superintendents attach much importance in their variable climate, is carried out to great perfection in the Utica Asylum. In fact, this was the first American

---

* As is well known, he has made sections of the brain and cord for microscopic examination and photography, covering whole areas of the actual size. He employs a very sharp knife 16 inches in length, to the ends of which upright handles are attached. With this formidable instrument he makes 400 sections to an inch, and can make as many as 500. The brain is hardened in alcohol and bichromate of ammonia. The amount of alcohol is gradually increased, and the fluid changed from week to week so long as it is turbid. Some brains require six or seven months' soaking. The brain is fixed in wax, or rather a composition of tallow, olive oil, and paraffin, in which material a brain was placed at the time of our visit.

asylum in which the system of forced ventilation and steam heating was introduced. A very striking feature of this, and some other asylums in the States, is the enormous number of visitors who come from motives of curiosity to visit the institution. Last year as many as 6,881 passed through the wards, in addition to 4,977 friends of patients. This must be a great inconvenience to the officers; at the same time it may have its advantages in giving confidence to the public. And yet, strange to say, there exists in regard to the management of asylums a very uneasy and suspicious feeling, from which Utica has certainly not escaped, notwithstanding the long service and celebrity of Dr. Gray. In the last annual Report of the managers, reference is made to this unfortunate feeling, and it is observed: "The actual accessibility of these institutions to all who wish to be informed of their management, and their actual inspection by thousands of visitors annually, ought to make such persons as are perfect strangers to them hesitate to insinuate charges against them except upon evidence that is verifiable and indisputable. . . . The officers of State asylums, whether managers or medical men in immediate charge of patients, can have no possible interest in retaining those who have recovered, but every interest in their discharge as soon as their condition will warrant it. . . . The report of every year contains the record of a number of persons who are brought to the asylum as insane on legal papers approved by courts, and believed to be insane by those committing them, but whom, upon examination and observation, the superintendent and medical officers found to be sane and so discharged them. This year there are 13, and during the history of the institution there have been brought as insane and found by the officers to be sane, and so discharged, 251 persons."

While this statement exonerates the asylum from receiving sane persons, it cannot but strike an Englishman as very remarkable and unsatisfactory that so large a number of sane persons should be certified to be insane who are not so, making allowance for a certain number of cases in which there may be a legitimate difference of opinion. It is the more remarkable, seeing that these certificates have received the sanction of Courts of Law. The conclusion seems to be inevitable that, in the State of New York at least, medical men sign certificates of lunacy who know very little about it, and that Courts endorse these documents as a matter of form, and that, therefore, this endorsement is not the check which the public no doubt expects it to be.

Dr. Gray holds decided views on many points, and, as is well known, is a stout defender of the use of mechanical restraint, including the crib-bed. He does not, however, approve of seclusion, and prefers a waist-belt and chair for an excited patient. The record of restraint was freely placed at my disposal. The month's entries showed that 14 patients had been restrained for periods of time varying from one day to four weeks. Restraint took the form of belt and wristlets, jacket with blind sleeves, leather muff, and the covered bed. Of these we saw examples as we passed through the wards.

A point to which Dr. Gray attaches much importance is classification. He holds that asylums should generally have as many as 12 wards for each sex, and as many as 14 if the number of patients amounts to 600. He observes that the collection of great numbers in one ward "destroys all idea or semblance of home or domestic natural life. . . A great dormitory-system is simply a process of herding people together, and an institution becomes a mere barracks and all individuality is lost." He has, therefore, no affection for the scene presented by some of our English asylums constructed to accommodate large numbers of pauper lunatics.

Dr. Gray is one of the few superintendents who have made special arrangements for the supervision by night of the acutely suicidal patients. On the women's side there were 20 of this class in one ward, and 30 slept there under the supervision of a nurse sitting by the open door. The night-watch, who visits every hour, is a further check on anything happening.

There are at the Utica Asylum four assistant medical officers, of whom one, Dr. Brush,* has visited the asylums of this country. Another, Dr. Blumer, is himself an Englishman.

This is a State institution (New York) under a Board of Managers, nine in number. These appoint the superintendent, and his subordinate officers on his nomination. The Board meets quarterly to transact business. A House Committee meets much oftener. Visitation is effected by (1) One or more members of the Board of Management; (2) The State Commissioner of Lunacy, Dr. Stephen Smith, who is appointed by the Governor and Senate, visits from three to five times a year, and reports to the Governor. Further (3) the Board of State Charities, consisting of 11 members and four ex-officio members, inspects the institution through the sub-committee appointed for asylums and almshouses. This Board is appointed by the Governor and Senate, and visits the asylum twice a year.

The number of patients is 600, namely, 309 males and 291 females. They are of various classes.

As to the class of patients mentally, chronic cases of insanity ought, as a general rule, to be sent to the Willard Asylum, so allowing more room for acute cases at Utica, but the admissions are not determined by the asylum authorities. The County Boards of Supervisors frequently send cases to neither Willard nor Utica, but to almshouses, because the cost is less. There were six cases of general paralysis among the women, and 32 among the men. The highest number of epileptics at one time during the last six years was 22.

* Dr. Brush has recently been appointed Deputy Medical Superintendent of the male departmnent of the Pennsylvania Hospital for the Insane, Philadelphia, an appointment on which all concerned with the institution may be congratulated.

The cost per week is 24s. With regard to the class of patients socially, and the charges, one-fifth to one-fourth of the inmates are pay patients, the amount varying from 24s. to £2 a week. If they have a special attendant and separate bedroom, patients pay from £4 to £6 a week. Last year £8,600 was received from private patients. The rest are paid for by the counties at the rate of 16s. 8d. a week. The county pays for clothing and incidental expenses, which average about 1s. 9d. weekly.

The proportion of attendants to patients is creditable, namely, 1 to 7. Their salaries are, on the male side of the house, from £4 to £8 per month. The female supervisor gets £5 a month.

With regard to recoveries, calculated on the admissions, the percentage varies in different years from 19·20 to 56·07. Last year it was 31·06. The total number of admissions from the opening in 1843 to 1883, was 15,267, of whom 5,508, or 36 per cent., were discharged recovered.

As to deaths during the same period, the percentage upon the average population varied from 5·70 to 18·14. Last year it was 9·64.

There is at this asylum a Turkish bath for the men patients, but it is not quite up to the mark. The ordinary bath-rooms are very good. As to medical treatment generally, I do not find anything specially to comment upon. About 30 women were taking sedative draughts at bed-time.

I find the average percentage of men employed to be 35·31; women, 37·93. In September, 1883, 67 men and 47 women were employed in general work; 16 men and 27 women in ward work; 27 men and 43 women in dining-room work; being a total of 110 men and 116 women employed in a daily average population of 606 patients. At the time of our visit, 105 men were occupied, namely, 57 in-doors and 48 out of doors; on fine days 60 would probably be engaged in outdoor work; 106 women were employed in sewing, and in the laundry, kitchen, and wards.

It is worth while giving here the regulations in regard to admission of patients into asylums in New York State, and the form of certificates. No patient can be confined as a lunatic in any institution in the State of New York without the certificate of two physicians under oath. Further, no person can be confined more than five days without such certificate being approved by a judge or justice of a Court of Record of the county. These may institute enquiries and take proofs as to any alleged lunacy, before approving or disapproving of such certificate, and may call a jury. The physicians signing the certificate must be of reputable character, graduates of some incorporated Medical College, permanent residents in the State, and in practice at least three years. The form of certificate is prescribed by the State Commissioner in Lunacy, and must bear date of not more than 10 days prior to admission. The form of medical certificate is as follows : "I, A. B., a resident of , being a graduate of , and having practised as a physician, three years, hereby certify under oath that on the day of , I personally examined C. D., of (here insert age, sex, married or single, and occupation), and that the said C. D. is insane, and a proper person for care and treatment under the provisions of chapter 446 of the Laws of 1874.

"I further certify that I have formed this opinion on the following grounds, viz. (facts). And I further declare that my qualifications as a medical examiner in lunacy have been duly attested and certified by (here insert the name of the judge granting such certificate)." Certificate of judge : "I hereby certify that A. B., of , is personally known to me as a reputable physician, and is possessed of the qualifications required by chapter 446 of the Laws of 1874, and I approve of the above certificate."

The next asylum visited was the *Massachusetts State Hospital for the Insane*, at *Danvers*. It has the character of having been one of the most expensive buildings for the insane in the United States. It was designed for 500 patients, and cost with the land over one million and a half dollars, or about £600 a bed, without furnishing. There are 200 acres of land. As might be expected, this large outlay has occasioned no small amount of adverse criticism, and much has been said in consequence against erecting asylums for the insane as if they were intended rather as palaces for princes than institutions for State lunatics. The tendency to extremes is extraordinary; and no contrast could possibly be greater than that presented by the cells of almshouses, from which many of the patients come, and the lofty rooms and spacious corridors provided by the State Hospital at Danvers.

The superintendent, Dr. Goldsmith, is intimately acquainted with British asylums. He is the youngest asylum superintendent in the States. He superintends with much vigour and kindness, and the condition of the asylum is throughout highly satisfactory, so far as management is concerned. The faults which exist have reference to the original construction and arrangement of the wards. It is remarkable that an institution so recently built should be provided with such obtrusive means of guarding the patients from the possibility of escape, and the windows from that of being broken.

One feature of this asylum is the appointment of a lady physician, Dr. Julia K. Cary, as assistant medical officer.

There are four assistant medical officers. Each has charge of two floors in the asylum. Dr. Julia Cary takes charge of the two lower floors on the women's side.

The government is vested in a Board of Trustees appointed by the Governor of the State. The number was originally five; two ladies have more recently been added. They serve five years, and are re-eligible. They meet monthly.

Visitation is effected by (1) members of the Board, monthly and oftener; one lady visits weekly; (2) The State Board of Health, Lunacy, &c., annually, and by their Inspector monthly; (3) The Governor of the State annually; (4) There is also a Board of Visiting Physicians.

The number of patients is 718; 337 men, 381 women. They are a mixed class socially. There are 106 private patients. Mentally, they are acute and chronic. The patients may be classified as the quiet, the more disturbed, the chronic and broken-down, the sick, and the refractory.

With regard to type of disease, there were about 40 patients, including six women, labouring under general paralysis. Dr. Goldsmith observes in his Report that the table of statistics of the causes of death shows that the number in any year is closely proportionate to the number of cases of general paralysis. Danvers is noteworthy for the large number of admissions. These amounted last year to 500, a large proportion being acute cases.

The cost per week is 15s. 8d. As to charge, nearly all private patients pay £1 a week, and have a bed and sitting-room. (Carriage drives are charged). Six pay £2 a week, four £3 a week, and have separate attendants. For a gentleman with an attendant the charge would be £4. The State pays 13s. 6d. a week for the poor patients.

The proportion of attendants to patients is 1 to 10, exclusive of the special attendants on private patients. Four men and six women are engaged as night-watches.

The salaries of attendants are as follows:—Males, £3 12s. a month, rising to £5 13s., the average being about £5. Women, £2 16s., with rise to £4, the average being £3 10s. No uniform allowed except caps.

With regard to restraint and seclusion, Dr. Goldsmith is strongly opposed to mechanical restraint except in very rare cases indeed. On the female side, restraint had only been resorted to during the last two years in cases of extreme violence. Several attendants were sitting by the side of, and endeavouring to soothe, noisy and excited women. During last year, only one man had been in restraint.

Danvers is the only hospital for the insane, among those I visited, which was provided with padded rooms, of which there are three. One man was restrained. Seclusion is but little used, as the records showed.

The recoveries given in the tables are for the past year only, and so in regard to deaths. They do not afford any information as to the percentage of deaths on the mean number resident, except for the same period.

As to employment, a number of patients are employed in brush-making, which takes the place of sewing among the women. It is easy work, and has been found very useful. It is stated that 40 per cent. of the men, and 43 per cent. of the women, were employed in one way or another during last year. This includes hair-picking, work on the farm, and in the laundry, kitchen, &c.

It may be stated here that there is a (locked) letter-box placed in the wards for the use of patients, as at Northampton, &c., which the inspector from the Board of Health, Lunacy, and Charity examines every month. As a rule he finds very few letters.

In consequence of the overcrowding in this and other asylums in Massachusetts, Dr. Goldsmith suggests the boarding-out of a select number of patients in private families, accompanied by systematic supervision by a properly qualified medical man. The insane in large towns might, he thinks, be provided for in buildings of their own, as is already partially done. Demented and broken-down cases might be cared for in a department of the almshouses, provided they are subjected to vigilant inspection. But a large number would remain unprovided for, and for these it would become necessary for the Commonwealth to erect a building located near the present hospitals, though sufficiently distant to prevent the incurably demented habitually coming in contact with the curable. Such a building would be after the pattern of those at Willard or Middletown (Conn.), costing only about £60 a bed. Had the Danvers Asylum been old, Dr.

Goldsmith would have suggested its occupation by chronic cases, provision for curable patients being made in a smaller building in the vicinity.

The next institution visited was the *Worcester Chronic Asylum,* established by law in 1877 in the buildings of the old Worcester Hospital for the Insane, which was opened in 1833. The present superintendent, Dr. Quinby, was appointed in 1879; he is the Secretary to the New England Psychological Society. The well-known Dr. Woodward was the first superintendent of the original hospital in 1833.

What is really remarkable in this asylum is, that an old building in the town, very near the Railway Station, is rendered exceedingly comfortable and homeish by the ability and attention of the superintendent, who devotes himself to making the wards and the patients clean, neat, and comfortable to an extent which is highly creditable. In fact, with such minute and considerate attention to details, I can well understand how patients find themselves more at home in such a building, whatever its structural defects may be, than in asylums with wide corridors and lofty day-rooms, built on the most expensive scale.

There is one assistant medical officer. The government is vested in a Board of Trustees, five in number. Two women trustees have been recently added.

Visitation is made by the State Board of Health, etc., at least once a year; also by the governor and council annually. The inspector (Mr. Sanborn) visits almost monthly for the above Board.

The number of patients is 387; 209 men, 178 women. There are none of the private class. The weekly cost is 12s 8d. The State pays 13s. 6d., which includes clothing. The town pays extra for clothing and breakage.

The salaries of attendants are as follows:—Supervisor, £4 per month; men attendants begin with £4, rising to

£4 12s. per month; nurses begin with £2 12s., rising to £3. Dr. Quinby said that the chief difficulty lay in getting good men. He speaks of the pay as little better than that of a day labourer.

With regard to restraint and seclusion, I found that since January 1st, 1884, no mechanical restraint had been resorted to on the male side of the house, except in the case of one man who had worn a belt and wristlets, on account of his attempts to commit dangerous acts. He nearly succeeded in killing one attendant and two patients. Another patient was restrained in the same way at the time of my visit, but was very loosely fastened. Seclusion had been employed since January on only three occasions, and for a short time. On the female side there had been, during six months, viz., January to June, three women on an average in some form of restraint, and there was a continuous record of seclusion in regard to one woman. During the three months prior to my visit, there had not been more than two women in restraint at any one time. Crowding has been the chief cause of restraint, it not being thought safe to allow certain patients who sleep in the corridors to be free. There is no crib bed. Dr. Quinby characterized it as a "nasty arrangement."

No recoveries are reported. There were during last year eight discharges, four having been removed to poorhouses, three taken by friends, and one male having "eloped," which, *more Americano*, means nothing worse than "escaped." The statistical tables in the Report do not afford any information as to the recoveries or the mortality of patients during the history of the hospital. As is observed in the annual Report: "From the nature of the cases of disease, no very satisfactory results can be expected." If, however, curative treatment is out of the question, the patients are made as comfortable as is possible.

Lastly, as to employment, it is observed in the Annual Report (1884): "Although we compel no one to work, we use every effort to persuade patients to do so, however little their labour may be worth. As a matter of fact, it is worth but little save to the patient himself. A record of the number of day's work performed would be misleading as, save in very exceptional cases, it would not mean a days' work in the ordinary sense, but simply that the patient had been more or less employed during the time specified; nor would it necessarily mean that anything had been added thereby to the income of the asylum, for in a majority of cases it actually costs more for necessary supervision than the work itself is worth."

The *State Lunatic Asylum* at Worcester is some distance from the town, and bears a general resemblance to that at Danvers. It is superintended by Dr. Park. It was opened in 1877, and was designed for 500 patients. There are 300 acres of land. The cost per bed was about £250. The view from the front is very extensive and beautiful, comprising Lake Quinsigamond, seven miles in length, which is frozen over in winter. The centre of the building contains the offices of administration; the wings on either side admit of complete separation of different classes, and retreat *en echelon*. When shown into the reception-room, we found several persons waiting to be shown round the establishment, according to American custom. Visitors arrive every day except Sunday. This asylum is characterized by the excellence of its management, as well as by the size of the building, the spaciousness of its rooms, and the handsome way in which many of them are furnished.

There are four assistant medical officers. A lady physician will be appointed this year.

The Board of Trustees consists of five men and two women, who meet once a month. Visitation is made by the State Board of Health, Lunacy, and Charity.

There are 749 patients, of whom 378 are males.

The patients belong socially to the private and public classes: State patients, 169; town patients, 461; private, 119. Although the old Hospital is reserved for incurable cases, the new institution admitted last year 65 chronic maniacs, 28 cases of chronic melancholia, 9 cases of chronic dementia, and 17 of senile dementia. There are 30 criminal lunatics in the house—a great misfortune. The average cost per week is 14s 5d., and the charge per week 13s. 6d. Some patients pay from £3 to £7. The proportion of attendants is 1 to 10.

As regards restraint and seclusion, several criminal patients were in restraint. Some are of a very dangerous class, and ought to be in a special asylum for criminal lunatics. The total number of patients restrained was 16. Three of the criminal patients are in constant seclusion.

Of the total number of cases admitted since the opening of the asylum, viz., 13,487, 4,857 recovered, or 36 per cent.

The annual percentage of deaths on the average numbers has varied from 3·73 to 15·90. During the last six years it has varied from 7·29 to 9·32. Since the opening of the institution to the present time, the percentage of deaths on the mean number resident was 9 per cent.

In his treatment, Dr. Park uses hyoscyamine; $\frac{1}{6}$ of a grain is the largest amount he has given hypodermically. He usually gives $\frac{1}{10}$ or $\frac{1}{12}$ of a grain. He does not often give it by the mouth. In several cases he has had to administer an antidote. He only gives it when the patient is in bed or in a chair.

In reference to employment, it is stated in the Annual Report that "a large number of patients in charge of attendants have been employed in outdoor labour with much benefit to themselves." On the women's side, "a large number who would otherwise have spent their time in absolute idleness or worse, came to be quite expert in the use of knitting-needles in consequence of the purchase of yarn in order to encourage this work." Hand-looms for weaving rugs and mats were being made for the use of both sexes.

The *Butler Hospital for the Insane,* at Providence, Rhode Island, was next visited. Special interest attaches to this institution from its having been, for more than 20 years, superintended by the late distinguished alienist Dr. Ray, through whom it has become familiar to English physicians. Fortunately, the present superintendent, Dr. Sawyer, has carried on the administration of the institution in the same spirit as his predecessor.

A very pleasing feature of this hospital is a drive in the grounds of the institution, $2\frac{1}{2}$ miles in length, through beautiful trees and by several streams of water. I do not know any asylum for the insane possessing within its own boundary as fine a drive and more delightful views, although the hospital for the insane at Brattleboro possesses in some respects equal advantages. There are 150 acres, 50 or 60 of which are under cultivation. The nett profit of the farm last year was £550. On the farm are 25 cows, a large number of pigs, and 15 horses, 11 of which are for carriage use in taking out patients. Two of the female attendants drive. There are thirteen carriages.

An excellent additional building called "The Duncan," from the gentleman at whose expense it was built, must be mentioned among the features of this institution. It accommodates 30 patients, nearly all the bedrooms being single. The general condition of this hospital is highly satisfactory, and most of the wards are exceedingly well furnished, bright and cheerful. The bedrooms are built on one side of the corridor, and an endeavour has been made to break up the wards as much as possible, so as to avoid having too large a number of patients together.

"The Ray Hall" is so designated in honour of the late Dr. Ray, by whom it was planned. It contains a bowling alley, for the use of the female patients, 50ft. by 16ft. Another is provided for men. Above are the reading-rooms and museum. There is also a billiard-room.

The government of this institution is vested in a Board of twelve trustees.

As to visitation, in addition to the Visiting Committee, two of whom are appointed for weekly inspection, there is the State Board of Charities, consisting of nine persons, appointed by the Governor of Rhode Island, who himself visits occasionally.

The number of patients is 198; 92 males, 106 females. Socially they are of a mixed class. The cost per week is 34s. 6d.,* and the charge per week averages about the same, varying from 21s. to £10 a week. Forty-two patients are paid for by the town, at the rate of 21s. per week. There are 156 private patients, of whom twelve pay about £3; three pay £3 12s.; five pay £4; one pays £5; three pay £7; and one pays £10 per week.

The proportion of attendants to patients is one to between five and six. There are also five night-watches, four inside the building and one out.

Salaries of attendants: Male supervisor,† £180 a year; other male attendants, £5 5s. to £7 7s per month; Female supervisor, £100 a year; attendants, £2 18s. to £4 4s. a month.

With regard to restraint and seclusion, the record contained very few entries under either head. The use of mechanical restraint is quite exceptional. There were two women in seclusion in the refractory ward. In the door was a small slit, glazed, to allow of inspection. On the male side no patient was in seclusion or restraint.

In reference to recoveries and deaths, the only statistical table in the annual report, is one of admissions and discharges since the opening of the institution in 1848. 3,039 have been admitted, and of these 1,015 have recovered, being at the rate of 33·3 per cent. There have been 510 deaths. The mean annual mortality per cent. resident since 1848 is, as nearly as can be given, 10.

There is nothing special to remark in regard to the

---

* At Cheadle (Manchester Royal Hospital for the Insane), the cost per week is 42s.; at the York Retreat, 37s. 4d.; and at St. Andrew's Hospital for the Insane, Northampton (England), 40s. per week.

† Has been 25 years at the Asylum.

medicines used in this hospital. The common mixture of chloral (10 grs.) and bromide (15 grs.) is given at bedtime for insomnia, and repeated in the night if necessary.

The record of employment shows that there are 20 men employed out of doors and three indoors; while on the female side 26 patients are employed in the laundry, and in the Superintendent's house. "There is hardly a patient or employé who does not take some interest and find pleasure in some part of the farm operations." The recreation-room accommodates 180 patients and 20 attendants. An exhibition of the magic lantern is given on Tuesday evenings, and a concert or dramatic entertainment on Friday evenings, during nine months of the year. Dancing is not practised.

Certificates.—The medical certificate, signed by two physicians, does not waste words; it simply says:—"*We hereby certify that A. B., of C. D., is insane,*" and the order signed by a guardian or nearest relative or friend is equally simple: "*I request that the above named insane person may be admitted as a patient into the Butler Hospital for the insane.*"

The Secretary of State and the Agent of the Board of State Charities constitute a Board of Commissioners with whom any patient may communicate by letter (unopened), and whose duty it is to investigate complaints and correct abuses.

It may be added to the above summary that the Board of Trustees, in their report of 1884, state that insanity is evidently becoming more prevalent every year in the United States, and especially in New England. As the hospital is overcrowded, increased accommodation is to be provided; it is also intended to erect detached cottages.

Dr. Sawyer uses, for checking the night-watches, an electric clock, called "the American Watchman's Detector" (New York). The original price was £25. The extra expense of connecting it with the wards was about £16. A carpenter and a man from the office were engaged for a fortnight in the work. Altogether the clock cost the asylum some £50. Its advantages were laconically summarised by Dr. Sawyer as "simple, noiseless, and successful." It is a 25-station clock.

This asylum was planned by Dr. Bell, formerly well-known as the superintendent of the McLean Asylum.

In the public cemetery (Swan Point), not far from the institution, lie the remains of Dr. Ray, and on his tomb, which is of granite, is the following inscription:—"Isaac Ray, M.D., LL.D., born at Beverly, Mass., January 16, 1807. Died at Philadelphia, Pa., March 31, 1881. Superintendent and Physician of the Butler Hospital for the Insane, from January 1, 1846, to January 1, 1867."

There is also an inscription recording the death of his only son, B. Lincoln Ray, M.D., who died in 1879, aged 43, to the great grief of his father, whose end is said to have been hastened thereby.

After leaving the Butler Asylum, we proceeded, accompanied by the assistant medical officer, Dr. Hall, to the *State Asylum for the Incurable Insane*, Rhode Island. This institution, in operation about 12 years, is situated at Cranston, a few miles from Providence. Dr. Stimson is the deputy superintendent, under the direction of the Superintendent of State Institutions, Mr. Hunt. Taking all the institutions on what is called the "State Farm," namely the asylum, the prison, the almshouse, and the reformatory school, there was during last year an average of 1,138 inmates.

The government and inspection centre in the Board of State Charities and Corrections of Rhode Island. There are nine members, and they frequently meet at Cranston, and at Providence, where we had an interview with them. The chairman, Mr. Pendleton, kindly joined us at the farm, which consists of about 500 acres, eleven of which are available for the insane.

Of four buildings occupied by the male patients, one

is a cruciform building of wood on a stone foundation, uncellared, and of only one storey, with slate roof, for 72 mild cases, with two attendants. It cost about £1,200.

Another building of stone accommodates 12 patients, and is used as a hospital. It cost £1400.

A third building, 112ft. in length, providing accommodation for 46 patients, some of whom are epileptics, cost about £1,000. A fourth is designed for 50 patients, who are more excited. At one end of the corridor is an attendants' room, which commands a full view of the ward.

A building called the Women's Pavilion accommodates 55 patients. Only 38 patients sleep in this house, the remainder sleeping in the "Stone Hall."

This is the next building, and is 140ft. in length. It is for the quiet class, and provides 49 beds in single rooms. There is only one long day-room.

The next building is a hospital for 9 female patients.

A fourth house for female patients, similar to the one in the men's department in the form of a cross, provides beds for 60 patients.

The total number of patients is 316, the sexes being about equal. There are, it need hardly be said, no private patients. As to their mental state, although, as I have said, the asylum is for incurables, a recent law allows of acute cases also being admitted.*

The cost per week is low, namely, 9s. 4d., including salaries and clothing as well as food. A certain appropriation is made annually by the Legislature, and it often happens that a portion of it is returned to the Government. It may be added that the nett cost of all the inmates of the State Farm, sane and insane, during the year, was under £22 per head, while at the Butler Hospital the weekly cost was 33s. 9d.

The proportion of attendants is very low, namely, 1 to 15; in the ward for the quiet women, only 1 to 24. They stay a very short time, although well paid.

* This law has, I am glad to say, been repealed.

As to restraint and seclusion, I regret to say restraint is freely used. Wyman's apparatus for confining the patient securely to the bed was shown us. It consisted of shoulder-straps and belt, with a strap at the bottom of the bed to which the ankles were attached, and with this was connected a strap passing to the back of the patient, who was thus firmly secured to the bedstead. The "monkey jacket" is a canvas dress blind at the legs and arms. To this are fixed straps made of canvas, which go to the side of the bed. It was stated that there were 20 of these in a closet in the ward. Two women, in the Women's Pavilion, wore strait waistcoats, and were fastened to the bench on which they sat. Three women were attached to the bench, but without strait waistcoats. One was secured by the jacket, but was otherwise free. In another building a woman was secured to her seat, and in the hospital two females were restrained, one simply by a strap to her chair, the other not only by a strap but by the strait jacket. In the cross-shaped building for 60 patients, 6 were restrained.

No recoveries are recorded in the last report, which is not surprising, considering the class of patients. Beyond the fact that 46 patients died last year, no statistics of mortality are given.

As to treatment, it cannot be expected that medical treatment will be carried out to any considerable extent in an institution of this description.

With regard to employment, Mr. Hunt attaches great importance to labour, and endeavours to carry it out systematically, but the number able to work is very limited.

This establishment well exemplifies the advantages and disadvantages of this mode of providing for the insane poor. It is cheaply done; it is in good charge, and a very intelligent Board directs it. There is every facility for out-of-door work, and every desire to carry it out. At the same time the relative positions of the lay and medical elements are not calculated to work satisfactorily. Further, the large number of patients congregated in one room, and the insufficient staff of attendants, have led to undue personal restraint in order to protect quiet patients from the mischievous and excited. Such a system is only calculated for quiet, harmless lunatics.

From the State Farm, Rhode Island, we passed on to the *Middletown Hospital for the Insane*, Connecticut. Foremost among the characteristics of this Hospital must be placed the annexes which have been added to the main building, and the utilization of cottages in the immediate neighbourhood.

Of two annexes one, when finished, will be three storeys high, and will provide accommodation for chronic cases and a number of epileptics.

This building is of red brick with stone foundation, and accommodates 325 patients, 168 of whom are women, and 157 men. Dr. Kenniston and his wife reside in the house. It is provided with a good kitchen, the cooking being done by steam. The baking is done at the main building.

The corridor is 10ft. 6in. in width, 164ft. in length, and 10ft. in height. Ventilation is from open fire-flues and a ventilator. Heating is carried out by fire and by flue, with which are connected two grates in the corridor. The windows are guarded outside, but not inside. There are nine single rooms on the male and as many on the female side of the house. A roomy recess extends 58ft. from the wall, and is 52ft. wide. There is an attendants' room, and dormitories; woven wire bedsteads are in use here, as in so many American asylums.

In this annexe a housekeeper, and a farmer and his wife, reside, in addition to the assistant medical officer.

An infirmary building of two storeys has been erected at one end of the female ward, for 24 acute cases, in as many single rooms, arranged on one side of the corridor. By means of sliding doors two wards can be converted into six, so that separation can be effected very readily.

There are four assistant medical officers and a pathologist. Three, I am glad to say, were married. The Board of Trustees consists of twelve members appointed by the Senate of the State.

The trustees visit the asylum monthly by a committee; not a week passes without one member of the Board paying a visit. The State Board of Charities, which was organised about five years ago, frequently inspect the asylum. Their visits average one every fortnight in the winter. The land attached to the asylum covers 300 acres.

The cost of building, including land, was £200 per bed. A stone building for insane convicts cost £125 a bed.

The number of patients in this hospital is 892; 406 males, 486 females. In regard to the class of patients admitted it is satisfactory to find that out of 1,113 under treatment during the last twelve months not more than 11 belonged to the paying class, while 774 were paupers, and 328 were "indigent," that is, were paid for by the State and by their relatives. It is stated in last year's report that the amount received per week for these various classes was, with one exception, the same. They are all treated alike, unless urgent symptoms require special attendance.

The cost per week is 15s. 6d., which is the weekly charge for board, including washing, mending, and attendance, for those supported at the public charge. Curable and incurable patients are admitted. There are about 50 idiots, for whom Dr. Shew desires to have special wards provided. Out of 271 admissions last year 13 were general paralytics, two being females; only one woman had been previously admitted. From the opening of the institution in 1868 57 patients with general paralysis have been admitted. Dr. Shew has no doubt that there is a steadily increasing ratio of this form of insanity. There is a large number of epileptic patients (about 70). For these Dr. Shew strongly urges another building. Criminal lunatics are admitted, and occupy a separate house.

There are 93 attendants, or about 1 to 9. They are paid, per month:—Males, £4 to £6, and £9 for supervisor; females, £2 15s. to £3 12s.

With respect to restraint and seclusion, during last year ·5 per cent. of those treated were restrained. There is on each side of the house one "covered" bed. Dr. Shew mentioned the case of a maniacal woman who persisted in standing and was much exhausted, who, in his opinion, was greatly benefited by being obliged to lie down in one of these beds. At the time of our visit to the asylum no patient was restrained. Dr. Shew feels at liberty to resort to mechanical restraint as a *dernier ressort*. One patient was in seclusion.

The statistics of recovery show that from the opening of the hospital 611 out of 2,952 admissions have recovered, or 20·6 per cent. The annual percentage of deaths, calculated on the daily average number resident, ranges, since 1868, from 3·28 to 17·57. In 1882-83 it was 9·36.

In connection with treatment it may be stated that on the male side 13, and on the female side 17 patients were taking sedatives. Dr. Shew attaches much more importance to the prevention of insanity than to its medical treatment when it occurs.

More than 50 per cent. of the patients are employed. On examining the record of employment I found that 26 men were engaged on the farm, 18 in the stables, and 37 on the grounds. On the female side there were 24 working in the laundry, 10 in the kitchen, 28 in the sewing-room, and 23 sewing in the wards. The latter figure has the merit of being understated, as I saw a good many engaged in sewing in their rooms whose work was unrecorded. Dr. Shew in his last report thus expresses the wish that he had some useful outdoor employment for the female patients:—"Their work seems to be restricted to domestic labour indoors. Why not set apart a large garden plot for their especial use, and allow them to have the entire charge of it? The experiment is certainly worthy of trial. If successful, the advantages of such a project can scarcely be over-estimated."

To the foregoing it may be added that Dr. Shew has set aside a ward for male epileptics in the south hospital. Twenty patients are under the supervision of two attendants during the day, and under the care of a special attendant at night. There are two large dormitories separated by a corridor, in which the night attendant sits, the doors of the dormitories being always open to allow of constant observation.

Although I have pointed out the provision made by means of separate buildings as the most striking part of the management of the Middletown Asylum, I should observe that the arrangements made in the way of classification and treatment in the main building are admirable, and reflect great credit upon the zealous superintendent.

Passing on to the *Hartford Retreat*, I may say that to myself and to Dr. Baker the institution possessed more than ordinary interest, from having been designed to follow in the path of the York Retreat; and Dr. Butler, who was for about thirty years the superintendent, has faithfully carried out those principles of humanity and individualization which characterised the latter when so many other asylums were in an unsatisfactory condition.

I cannot do better than quote from a letter received from Dr. Butler since visiting him at his own house, in his honourable retirement from active practice.

"From the York Retreat and its founder, the Connecticut Retreat took, in 1824, both its name and guiding principle of treatment. It is a most grateful duty to one who for nearly thirty years was the superintendent of the latter, thus to recognise and assist the long-time quiet and unassuming source of this blessed power to heal the sick and alleviate the sufferings of the most grievously afflicted of mankind. That quiet, unpretentious hospital near York, and the influence for blessing which has gone out from it, recall what Emerson says: 'Every man is not so much a workman in the world as he is a suggestion that he should be.' And again: 'Men walk as prophecies of the next age.'

"As our poet says:—

> The hand that rounded Peter's Dome,
> And groined the aisles of Christian Rome,
> Wrought in a sad sincerity.
> * * * * *
> He builded better than he knew;
> The conscious stone to beauty grew.

"How cheering to us all to realise that this 'beauty' is being more and more recognised all round the world."

Dr. Butler retired from his office twelve years ago, and now resides at Hartford, at an advanced age, but takes undiminished interest in the great work of his useful life—the welfare of the insane. His annual reports form a valuable series of papers which, whether long or short, reflect the valuable experience of his long superintendency. Dr. Stearns is the worthy successor of Dr. Butler. The Retreat has been fortunate in its superintendents, those of former times having been Dr. Todd, Dr. Fuller, and Dr. Brigham.

The Hartford Retreat was at a very early period devoted exclusively to the insane, an asylum in Virginia having been the first, and the Friends' Asylum, at Frankford, having been the second to pursue this course. The Connecticut Medical Society originated the Retreat, and contributed 500 dollars to establish it.

The asylum stands in grounds resembling an English park. Eighty acres are devoted to the farm. The view from the balcony of the original building commands a beautiful prospect of the surrounding hills. The Connecticut river lies between them and the asylum, but is concealed from sight.

Dr. Stearns is justly proud of a very handsome villa on the grounds for a single patient, occupied by a lady.

There are two sitting-rooms, a bath-room, and four bedrooms. It is built of wood, with slate roof, the cost being about £800. It was presented to the institution. There are two assistant medical officers, one of whom, Dr. Page, recently visited our asylums in England.

This institution is under the care of directors chosen every year and a Board of Managers.

A visiting committee consisting of directors appointed for the purpose, and a committee of ladies, periodically visit the asylum. There are also six medical visitors.

There are 130 patients, the sexes being about equal. They are of the middle and higher classes. Both curable and incurable cases are admitted. Since 1869, when the institution was re-organized, 1,471 cases have been admitted; of these 395 laboured under acute mania and 263 under acute melancholia, 37 were cases of general paralysis, and 33 cases of senile insanity. The cost per week is £2 2s., including everything, and the weekly charge ranges from 24s. to £20. Three or four pay nothing. The patient in the villa pays £800 a year.

The salaries of attendants are, as a rule, £5 a month for males, and £3 for females; supervisor, £6 6s. a month.

Restraint is seldom resorted to. A crib-bed is used for a patient who would stand for hours during the night if allowed to do so.

Lastly, with respect to recoveries and deaths, of 6,488 cases admitted since the Retreat was opened 60 years ago, 2,890 have recovered, or 44 per cent. Of the above admissions, 793 patients died, but as the average number resident is not given, we are unable to state the percentage in a correct form.

After Hartford, New York was visited, where, as already intimated, every facility was given, through the instrumentality of Dr. Gray and Dr. S. Smith, for seeing the institutions for the insane, the first of which was the *New York City Asylum* for males (Ward's Island).

The unique feature of this institution is its salt-water swimming bath in the grounds of the asylum, to which I have already referred on a previous page.

Another feature is the large proportion of medical assistants to the number of patients, there being 1 to 100, some of whom, however, are unpaid, and are rather clinical clerks than assistant medical officers.

A third feature is the appointment of an assistant medical officer, whose duty it is to attend to the wards by night. He is on duty from 7 p.m. to 8 a.m., and patrols the wards twice during the night. This asylum is not, however, alone in this excellent arrangement.

It is stated in the Report of the State Commissioner in Lunacy, that the assistant medical officers take turns in such order that each will be on duty once in twelve nights, and so that an interval of at least one day elapses before the same physician returns to day duty.

The following, from the same source, illustrates the kind of report made each night by the medical officer in charge :—

10.30 p.m. Ward 11. Mr. S., being noisy, received medicine; quiet after.

11 p.m. Ward 14. Messrs. K. and B., being noisy, received medicine; quiet after.

11 p.m. Ward 21. Mr. B—r, being noisy, received medicine; quiet after.

11.30 p.m. Ward 1. Messrs. B. and K., being noisy, received medicine.

11.30 p.m. Ward 3. Mr. B—h, noisy, received medicine; quiet after.

11.30 p.m. Ward 9. Mr. B—y, noisy, had medicine; quiet after.

3.5 a.m. Ward 19. Mr. P., noisy, had medicine; quiet after.

3.5 a.m. Ward 9. Messrs. McK., and McN., restless, had hypodermics; quiet after.

3.30 a.m. Ward 15. Mr. B—n, noisy, had medicine as ordered; quiet afterwards. M. A. received stimulants as ordered during the night. All attendants in their rooms at 10 p.m.

Frequent inspection is secured by the visits of the State Commissioner in Lunacy (Dr. Stephen Smith), of the State Board of Charities, and also of three Lay Commissioners of Charities and Corrections.

There are 1,494 patients, all males, and of the poorer class.

The proportion of attendants to patients is 1 to 12.

As regards restraint and seclusion, the last entry of restraint was on September 10th, 1883, and the last entry of seclusion on the 18th September.* When a patient is much excited, the attendants have strict orders to send for the superintendent or one of the assistant physicians, in order that he may be promptly placed under medical supervision.

Since the year 1874, 4,611 patients have been admitted, of whom 692, or 15 per cent., have recovered; 810 have been discharged relieved, and 1,528 have died.

There is also on Ward's Island the *Emigrant Hospital for the Insane.* There are, in this asylum, 70 men and 60 women.

In one room in the female department seven patients were engaged in sewing. The more excited, 15 in number, were placed at the end of the building, and were in charge at the time of only one attendant. It was stated that this was temporary. It cannot be said with truth that these patients were in a satisfactory condition; in short, they were evidently neglected.

Twenty-five men were employed out of doors.

There was little satisfaction derived from visiting this institution, but there seems no good reason why, with the present building and the amount of land attached, the

---

* That is, more than a year before the date of our visit.

patients should not be made more comfortable, under vigorous and efficient management.

Bloomingdale Asylum, New York, was first visited, in the absence of the Medical Superintendent, Dr. Nichols, under the guidance of the senior assistant, Dr. Sanger Brown. On a subsequent occasion, I had the advantage of being Dr. Nichols's guest for several days. This asylum (a department of the New York Hospital) has been so frequently referred to in the foregoing pages, that a short notice will suffice. It is the oldest institution for the insane in the State of New York,[*] and the oldest but three in the United States. The New York Hospital was opened in 1791, and it so happened that the first cases admitted were insane patients. In 1808 a separate "Lunatic Asylum" was built by the Governors, which was not occupied till 1821, when the Bloomingdale Asylum was opened, situated in the avenue of that name. The great feature of this institution is the large amount of good it effects for the indigent classes who are socially above the poor. Thus, at the time of my visit about two-thirds of the patients were paying nothing, or else a sum below the cost of maintenance, the deficiency arising from these cases amounting, during the year, to more than £6,000. This deficiency was met in part by a charitable endowment—the "Green Fund"—and by the higher rates paid by the remaining third. Even then the expenditure exceeded the income by at least £1,600.

I was glad to find that, in summer, excursions are frequently made to the sea-side, or to a farm at "White Plains," twenty miles off. It appears from the report that four men and eight women were boarded in private families in the country, three of them being with the farmer at "White Plains" for several weeks, and in some instances for three months. This practice has been only

---

[*] Eleventh Annual Report of the State Commissioner in Lunacy, p. 326.

lately introduced, with, as may well be supposed, great advantage.

An important feature in the management of this Institution, and one deserving much more general adoption, is the large proportion of attendants to patients. For forty refractory patients there were no less than twenty attendants; so that, in addition to the advantage of ample supervision, patients labouring under paroxysms of excitement, whether by night or by day, can be thoroughly attended to, and the necessity for mechanical restraint reduced to a minimum or altogether avoided.

The visitor to this institution must be struck with the fine grounds, in which are seen pines, oleanders, and two very fine yew trees. There is a beautiful drive, two miles in length, round the estate, which comprises forty acres.

A recently erected addition for about eighty female patients, called the " Green Memorial Building," is in every way adapted to the purpose for which it is designed, and here, as at St. Elizabeth's, Washington, Dr. Nichols has displayed his architectural skill. The windows have large panes with ornamental frames outside, and command a fine view of the Hudson. This building contains a number of very good and well-furnished private rooms. Rugs are preferred to a carpet covering the whole floor, as being healthier and cleaner.

The case-books and journals of this institution are admirably kept.

There are two assistant medical officers. The number of patients is 247; 115 males, and 132 females. The cost per week is about £3 3s. for each patient. The weekly charges vary from £3 to £10. A special fund exists by means of which some patients can be treated, if curable, for twelve months, for £1 a week, or even gratis. A patient paying highly has the exclusive use of an attendant, daily carriage exercise, and wine.

With regard to the salaries of attendants, the highest pay is £6 for male, and £4 for female attendants.

No patient was in restraint or seclusion at the time of my visit. (See p. 55).

Recoveries.—Of 7,641 admissions from the opening of the asylum in 1821, 3,157, or 41·31 per cent., recovered.

Deaths.—Of the above admissions 1,036 died, but the average number resident during the period embraced is not stated, so that no percentage can be given. However, from 1872 (inclusive) the annual mortality has been about 9 per cent. of the average number resident.

The treatment pursued by Dr. Nichols has already been given (p. 59).

Before quitting the asylums of New York, I add the returns of the number of the insane and idiotic in this State, as given in the census of 1880.

	Insane.	Idiots.
In Asylums	8079	93
Institutions for Idiots		602
Almshouses	1577	537
Jails	6	1
Charitable Institutions	32	70
At Home	4361	4781
Total	14,055	6084

The population of the State of New York in 1880 was 5,052,871.

I proceed to notice the *New Jersey State Lunatic Asylum*, at Trenton. One of my objects in visiting it was to see Miss Dix (see p. 37), who has for many years of her life been accustomed to make it her home at leisure times. At an advanced age, and in very feeble health, she retains her interest in the chief object to which her energies have been devoted for more than fifty years. She referred with especial pleasure to her visit to Scotland in 1855, and the reform she was instrumental in effecting in the condition of the insane in that country.

This asylum, opened in 1848, is superintended by Dr.

Ward. Dr. Buttolph was the Superintendent for many years, and left for the asylum at Morris Plains* in the same State.

The institution, possessing or renting 350 acres, is pleasantly situated in the country several miles from Trenton, and presents a handsome exterior, including portico and dome. The centre building, four storeys high, has the steward's apartments, store-room, and kitchen on the first floor; parlour and public office on the second floor; and the Superintendent's private rooms and the chapel on the third floor. The fourth is occupied with bedrooms. The wings are three storeys high, and contain, as is usual in American asylums, a large number (300) of single rooms. With some exceptions, the rooms are placed on both sides of the corridors. Ventilation is secured by the customary fan.

There are two assistant medical officers.

The asylum is governed by a Board of Managers, consisting of 10 members, who report annually to the Governor of the State. They make monthly and weekly inspections by means of small committees. The State Board of Charities also inspects the asylum from time to time.

The number of patients is 674; 329 men, 345 women. The capacity of the asylum is for 500. It is therefore much crowded. The patients are chiefly of the middle-class; there are very few paupers. Since the Morris Plains† Asylum was opened, this institution is regarded as more especially for the curable class, but of the 177 admissions last year, as many as 36 were cases of chronic mania, and 33 of chronic dementia. Thirteen of last year's admissions had been insane from 5 to 10 years, five from 10 to 15 years, and as many as ten over 30 years.

The cost per week is 16s. 8d., and the average charge for

---

* Since the above was written Dr. Buttolph has retired.
† 297 patients have been transferred to the Asylum for chronic cases at Morris Plains.

paying patients, (from whom £125 per week is received), is 25s.

The proportion of attendants is 1 to 12; for the refractory class, 1 to 8. Their salaries vary, for males, from £4 10s. to £5 a month; females, from £3 5s. to £3 10s. The male supervisor receives £7 a month; the female supervisor £6.

In regard to restraint and seclusion, Dr. Ward prefers mechanical restraint to prolonged seclusion. Four women patients wore camisoles at the time of my visit, and in an airing court one wore a muff, and one was attached to the seat. On the male side, there was in one airing court, containing 90 patients with two attendants, one man restrained by a muff, and another by belt and wristlets. Several patients were in seclusion.

Of the whole number of cases admitted from the opening of the asylum, namely, 6,181, 2,179, or 35 per cent., recovered. The number of deaths was 1,199, but as unfortunately the average number resident is omitted, we are unable to give the percentage. It is greatly to be regretted that so few of the American asylum-reports afford this important information for the period of years for which the mortality is recorded.

With regard to employment, 50 men and 75 women are occupied in or out of doors. In mid-winter only 15 men could be employed out of doors. On the female side a satisfactory amount of work is executed in the sewing-room. Last year this comprised 370 dresses, 204 chemises, 272 vests, 219 drawers, 119 pairs of socks, and 132 curtains. There is a large room designed for calisthenics, but it is chiefly used as a sewing-room. A bowling-alley for women—an unusual feature of an asylum—is provided at this institution. There are frequent exhibitions of the stereopticon; and entertainments consisting of dancing, theatricals, &c. The patients are well supplied with books, of which there are about 2,000 in the asylum library. The pleasant circumstance occurred here of a former female attendant presenting a library, which is called after her name.

Certificates.—A request for admission is made by a friend of the patient, accompanied by a certificate of insanity from a respectable physician; and a bond with satisfactory sureties is required. A judge of the Court of Common Pleas gives an order for admission, after calling to his aid a physician and other witnesses. If he deem it necessary he can call a jury. The form of medical certificate is as follows:—" I, A. B., physician, of the township   , in the county of   , do hereby certify that I have examined into, and I am acquainted with, the state of health and mental condition of C. D., &c., and that he is in my opinion insane and a fit subject to be sent to the State Lunatic Asylum."

In New Jersey I also visited the *Hudson County Lunatic Asylum* (Jersey City). It contains 246 patients, 89 being males, and 157 females. It has a capacity for only 150, and is therefore excessively crowded. It was built 11 years ago. A separate building for excited patients was built two years ago. In the new building are strong rooms, with a narrow space between them for the purpose of ventilation. There are six rooms on each side of this space. The doors have bolts above and below, and a wicket with a wooden bar. The rooms are lighted from above the doors. The floors are stone. There is a wood bench attached to the wall, on which a patient who destroys his bedding can lie, and escape the cold floor. There were three women in seclusion in the rooms on one side; and on the other there was a youth in the first room, a woman in the second, and a man in the third.

A building has just been erected for the excited men, one of whom was in the corridor restrained by a muff, and suffering from lupus. There were several men in a small airing court adjoining. None were in seclusion.

There are eight male and nine female attendants, or about 1 in 14.

As the asylum at Morris Plains, superintended by Dr. Buttolph, is for the chronic insane in New Jersey, and as it is not full, it seems surprising that some of the patients now in the Jersey City Asylum are not transferred thither; but the explanation, no doubt, lies in the fact that while the weekly charge at the former is £1 a week, it is only 12s. at the latter, and of this 4s. is paid by the State.

It is to be regretted that the attempt has been made to provide such accommodation as that now described, which consists of bare cells. It cannot have been done under the direction of any medical man acquainted with modern asylum construction. The New Jersey City authorities cannot be complimented on their action in this matter, which seems to have been mainly directed by one motive—economy.

A young physician is in charge, and very courteously showed Dr. Nichols and myself over the establishment.

The census of 1880 showed that there were in this State, the following number of insane and idiots:—

	Insane.	Idiots.
In Asylums	1632	56
Almshouses	116	127
Jails	3	
Charitable Institutions	2	1
At Home	652	872
Total	2405	1056

The population of New Jersey was 1,131,116 in 1880.

An object of special attraction for psychological travellers in the States is the *Pennsylvania Hospital for the Insane*. This institution, popularly known as "*Kirkbride's,*" situated in the vicinity of Philadelphia, will be always associated with the name of the distinguished physician, who for forty years superintended it with so much zeal, ability, and kindness. His memory will ever be treasured by those who knew him, whether patients or others, and when a generation has sprung up that knew him not, he will still live in the beneficent results of his long life's labour and earnest work. It was not to be expected that at the age he had attained, his benign reign should last much longer. He is dead, but *non omnis moriar* is emphatically true of him, when his life and its effects are considered. Dr. Kirkbride passed away in December, 1883, and Dr. Chapin reigns in his stead.

Dr. Kirkbride in his sketch of this Hospital, written in 1846, after observing that " from the York Retreat soon after its foundation emanated a code of moral treatment which even at this day can hardly be surpassed," states that " the mild and rational system pursued at the Retreat was soon adopted in the Pennsylvania Hospital for the Insane; long, indeed, before a reform was more than thought of, in many of the establishments of a similar kind in Great Britain."

This institution is a department of the General Hospital in Philadelphia. There are two distinct buildings, one for the male, and the other for the female patients. Dr. Kirkbride held strongly that this complete separation of the sexes is desirable.

As I have already spoken of Dr. Kirkbride's opinions in regard to classification, it is unnecessary to refer to them again.

It is difficult to single out any one feature in the management of the Kirkbride Asylum, but perhaps the point which he would have more especially dwelt upon with satisfaction, was the social meeting of the patients on very frequent occasions for recreation, readings, and amusements of various kinds, in which he himself took part for so many years with unflagging interest. Especially did he take an interest in, and attach importance to, the officers' weekly tea-party—established in 1866, and continued from that time. On these occasions about 40 persons, three-fourths of whom were patients, were invited from every ward in succession, to meet at the officers' table, Dr. Kirkbride himself presiding at this meal. Including this tea-party, every evening of the week during nine months of the year had its appropriate engagement.

The institution is managed by a Board of Trustees, consisting of 12 members. It is visited every fortnight by members of the Board. The Lunacy Committee of the State Board of Charities inspects periodically.

The cost of building was £280 per bed, the number of patients being 368; males 175, females 193. The number of acres is 113.

The charge per week averages £5. The weekly cost is about 28s., exclusive of clothing.

The proportion of attendants is at least one to three. Their salaries are:—Males, £3 15s. to £4 4s.; female supervisor, £5; lady companions, £6; and female attendants, £2 18s. a month. There are three lady companions on the female side.

With regard to restraint and seclusion, the muff is occasionally employed for very violent cases. As is well known, the late Dr. Kirkbride was distinctly of opinion that mechanical restraint was advantageous in certain cases. It is stated that there was an average of two patients restrained last year. On September 30, 1884, one male patient was restrained by wristlets, for destructive habits and denuding the person. Two violent male patients were in seclusion at the same time. Two women were restrained, and two secluded.

Dr. Chapin, the successor to Dr. Kirkbride (see p. 77), has recorded his views on treatment in the second Report of the Committee on Lunacy, of which I avail myself. He uses warm baths, with cold applications to the head, in certain cases of acute mania, &c. The cold bath, douche, and shower bath are not used. Counter-irritation is rarely employed. Opium is occasionally prescribed, as in some forms of melancholia, with wakefulness, dilated pupils, and a soft compressible pulse. He attributes rapid dementia to its prolonged administration. Digitalis is chiefly given in the noisy stage of general paralysis. Of hyoscyamus Dr. Chapin observes that the contradictory results obtained from the use of this drug are due to its different habitats. It appears that while the leaves and preparations of the English plant are reliable, those of the American plant are comparatively inert. Chloral is occasionally used in acute delirious mania, and in the *status epilepticus*. Bromide of potassium (or, in preference, bromide of sodium) is used with most excellent results in mania and in insomnia, as well as in epilepsy. Dr. Chapin is fond of the combination of bromide, tincture of hyoscyamus, and the fluid extract of cannabis indica. Hyoscyamine (Merck's white crystals) is administered, by mouth or skin, in states of mania with excessive motor restlessness.

I have already referred to the employment of some of the female patients in pottery work of a useful and ornamental kind (see page 63).

There is a gymnastic hall, which is used for two evenings in each week in winter, for calisthenics, skittles, dumb-bells, and music. No patients are allowed to do any menial work in the wards.

Without entering upon any details of construction, it may be well to note that the department for males, situated in fifty acres of pleasure grounds, consists of a centre building with wings running north and south, the frontage extending 512 feet; also other wings running east, 180 feet; all being three storeys high; the last communicate at their extreme ends with one-storied buildings for the excited and noisy—an arrangement to which Dr. Kirkbride attached great importance. The airing courts connected with these have an open palisade in front. The walls, of stone, are stuccoed, the interior being brick. The centre is 115 by 173 feet, and has a Doric portico of granite; it is surmounted by a handsome dome. For the 250 patients for which this building provides, there are fourteen wards. Ventilation is secured by the customary fan; heat, by fresh air passing over steam-pipes in basement, the temperature being 70° in the winter; lighting, by gas. There are 220 rooms 9 feet by 11 feet, and 20 rooms 15 feet by 12 feet; there are no associated dormitories. Infirmaries are not provided, but there is a special ward to which patients when ill can be removed. Epileptics are not separated.

The building for the female patients presents also a centre and wings, with a frontage of 436 feet. The former, like the building for the male patients, has a Doric portico, and is surmounted by a dome. There are about 50 acres of land, 20 of which are gardens, and 25 are appropriated to the recreation of the inmates, one hour in the morning and one in the afternoon. The particulars given in regard to the department for the males apply also to this building.*

It is greatly to the credit of the management of this excellent institution that, with the exception of an appropriation by the Provincial Assembly of £16,000 towards the original hospital buildings in the city, the work has been carried on by means of subscriptions, no aid having been sought from the State, county, or city.

It may be stated that Dr. Kirkbride,† whose life was devoted to the insane, was first connected with the Hospital in 1833, as a Resident Physician in "The Pennsylvania Hospital" in Pine Street, Philadelphia, which was established in 1751, and opened in 1752.‡ In 1841 the Insane Department was removed to its present locality in West Philadelphia, at which time Dr. Kirkbride was elected the Medical Superintendent. There was then only one hospital for both sexes, but in 1859, a new one was opened

---

* See 2nd Report of the Committee on Lunacy of the Board of Public Charities of the State of Pennsylvania, 1884, which contains a useful description of the hospital.

† A memorial of Dr. Kirkbride, condensed by the writer, will be found in the "Journal of Mental Science," July, 1884.

‡ In 1796, the insane in the old Hospital, who had up to that year been confined in cells in the basement, were removed to the west wing, then completed. Here they remained until the present hospital for the insane was opened, two miles west of the Schuylkill river, in 1841.

for male patients, the females occupying the original building. Both were superintended by Dr. Kirkbride. They are a third of a mile apart, and have a capacity for 250 patients each.

During the period from 1841 to 1883 the admissions into the Pennsylvania Hospital for the Insane amounted to 8,852. Of these 3,958 were discharged recovered; 680 much improved; 1,509 improved; 1,156 were stationary; and 1,167 died. No information is afforded as to the relapses.

The provision made for the insane and idiotic in Pennsylvania is shown by the following figures from the census of 1880:—

	Insane.	Idiots.
In Asylums	3299	65
Almshouses	1181	468
Jails	3	
Institutions for Idiots		323
Charitable Institutions	4	8
At Home	3817	5633
Total	8304	6497

The population of Pennsylvania in 1880 was 4,282,891.

The next asylum visited was *Norristown*, a few miles from Philadelphia. This institution was erected to provide, in some degree at least, for the accumulation of chronic cases in the almshouses of the South-Eastern District of Pennsylvania, and it now constitutes for this district the State Hospital for the Insane. There is a County Visiting Committee appointed by the Lunacy Commissioners, consisting of three men and three women. Dr. Chase is the Resident Physician in the building devoted to the male patients, and Dr. Alice Bennett is the Resident Physician in the building for females. There are two male and two female assistant physicians.

There are four separate buildings, connected by open

corridors, 100 feet in length. There are four wards in each block, and they are called after the letters A B C D. There are collected in the D ward of each building all the cases of the three other wards which require attention during the night, such as the sick, the epileptic, the suicidal, the restless, &c. By this means, about 25 in each block are subjected to close surveillance. It is especially noteworthy that patients are employed as night-nurses. Thus out of 13 men employed at night eight are patients, and receive no remuneration. Of this force a Captain of the Watch takes charge, and they go through the wards once every hour. A nurse and two patients remain in each section.

On the female side a new ward building has recently been occupied, greatly to the relief, it is stated, of the excited patients, for whom it is designed. It so happened that on the occasion of my passing through the wards there was a large amount of excitement. It is probable that, could the women be employed like the men in outdoor work, the amount of excitement would be materially reduced. Within a recent time two infirmaries for different stages of illness have been completed and placed under skilled nurses, along with a night-watch. This is spoken of as having been of the greatest possible advantage.

The government is vested in a Board of Trustees, 13 in number, and visitation is effected by members of the Board, and by the Committee on Lunacy of the Board of Public Charities.

The cost of building was £160 per bed. For new buildings, which are designed to be one storey in height, with kitchen and dining-rooms, the estimated cost is £80 per bed.

The number of patients is 1,110; 572 males, and 538 females.

Of 1,032 patients, the average during last year, 970 were public and 62 were private patients.

As regards the class of patients mentally, since the opening of the institution 406 cases of acute mania have been admitted, 534 cases of chronic mania, 310 cases of acute melancholia, 109 cases of chronic melancholia, 461 cases of chronic or senile dementia, 112 of epilepsy, 69 of idiocy or imbecility, and 53 of general paralysis. From the above it will be seen that this asylum has a considerable number of acute as well as chronic cases.

Cost per week.—14s. 2d. including clothing. It may be stated that it is required by law that the cost of public patients shall not exceed 16s. 8d. a week.

The proportion of attendants to patients is 1 to 10.

Restraint and Seclusion.—Mechanical restraint is avoided. Dr. Chase observes in his report to the Committee on Lunacy: "I am pleased to be able to report that we have no system of restraint or coercion in use in the hospital. Under the beneficial effects of good food and occupation, the patients are quiet and orderly to a remarkable degree, and the cases of excitement are either so transitory or mild in nature that the simplest of means are sufficient to control them." And speaking of the female department, Dr. Alice Bennett says: "No kind of mechanical restraint is employed. Seclusion is only occasional, and always temporary. Violent patients are distributed in large wards, and given as much out-of-door exercise as practicable, and supervised by a larger force of attendants than the milder cases. ' When the propensity to tearing clothing exists, dresses made of canvas are sometimes substituted temporarily for the ordinary dress."

Recoveries.—Out of 2,234 patients admitted since the opening (July, 1880) 383, or 17 per cent., have recovered.

Deaths.—406 have died, but as the average number resident is given for the women only, no aggregate calculation of any value can be made. With them it was 9 p.c.

Treatment.—Dr. Chase orders the warm bath at bed-time from a quarter to half an hour, with cold to the head, in cases of excitement and sleeplessness, the temperature of the bath never being higher than 98 degrees. He frequently uses local blood-letting. He rarely uses counter irritation. In exhausted maniacal states he gives freely milk, eggs, and whisky stirred together. For opium he has not much liking, and thinks that of late years the cases of insanity are not so well suited for this drug. He regards hyoscyamine as so dangerous a drug, in even moderate doses, that he has entirely abandoned its use after a long trial, and much prefers the old-fashioned hyoscyamus. As to chloral, he favours its use as possessing the good qualities of opium without its objectionable features. Next to chloral he ranks bromide of potassium, but he finds it does little good unless given in full doses and continued for a long time. Paraldehyde is a favourite.

Dr. Alice Bennett has paid much attention to gynæcology, and has made a number of observations which, when still further extended, will doubtless be of much value, to whatever conclusion they may lead. It is stated that the number of patients benefited by treatment for uterine disorder during last year was 60 out of 125 so treated.

The number taking hypnotics at bed-time during last year averaged six.

Employment.—I have already spoken of Norristown as an asylum in which a considerable number of patients are employed. I may here state that 150 patients were employed during last year in improvements on the grounds; six were engaged on the farm, and eleven in the garden, while indoors four were occupied in the printing office, nine in the saw-shop, and five as tailors. Dr. Chase strikingly states that the advantage of labour has been nowhere more shown than among the violent patients, of whom about 70 per cent. work out of doors.

Among amusements may be enumerated games and music, reading, lectures, magic-lantern exhibitions, concerts and theatricals, and outdoor recreation, as lawn tennis, football, croquet, fishing, and picnics. There is also a bowling-alley, and a billiard-room.

While in Philadelphia I visited the *Blockley Almshouses*, where there are 718 insane inmates. Dr. Richardson is the Superintendent, and Dr. Henry the assistant medical officer, both of whom showed me all that was possible in a hurried visit which I paid to the institution. It was built 53 years ago, and is totally unsuited for the purpose to which it was devoted. A more recent building only increases the dismal condition of the house by lessening the available space. Some of the windows are also darkened by a provision recently made for escape in case of fire.

There were about 80 patients of the excitable class. I saw three men in seclusion, one of whom was confined by a leathern muff. One woman in seclusion wore a belt and wristlets. For suicidal cases restraint is frequently used, especially during the night. In the latter case the watch looks into the room from time to time. The strong-rooms have grated shutters and iron bars, and are whitewashed. The door, which has a wicket in the centre, is fastened by bolts. One of the men I have mentioned had a bucket to sit upon, while others sat all day on the floor, on which, in some instances, was a straw bed. The dietary was fair.

The resident medical officers are in no way blameworthy for the state of things existing in this almshouse or hospital, but, on the contrary, do what in them lies for the patients entrusted to their care. The whole place is, as I have said, totally unsuitable for the custody, to say nothing of the treatment, of the insane, and should be pulled down.*

The Frankford Asylum, founded by the Society of Friends, is situated a mile from Frankford, which is a ward of the city of Philadelphia. It was projected in 1813, and opened in 1817 for the reception of patients.

---

* A fire occurred some months afterwards, which partially destroyed the building, with, unfortunately, some loss of life.

In the first Report it is observed that "as this extensive edifice was to be adapted to a plan of restorative treatment in diseases of the mind, on a system hitherto without example in this country, it was to be expected that numerous difficulties would present themselves, requiring much consideration and time to surmount. The work has nevertheless been accomplished, with the hope at least, on the part of those to whose care it was intrusted, that it may answer the benevolent intention of the individuals who so liberally contributed the pecuniary means to effect it."

The buildings consist of a centre and wings, with but little attempt at ornament. The window sashes are of iron, the upper half unglazed with moveable wood sash. The house is heated by steam, and ventilated by the fan. The walls of the corridors and rooms are relieved by pictures. There are 85 single rooms. The wards for the most excited patients consist of one storey.

Dr. Hall, the Medical Superintendent, succeeded Dr. Worthington, now retired into private life, whose name was well known for many years in connection with this institution. There is one assistant medical officer.

The government is entrusted to a Board of 20 Managers. A Committee of the Board pays weekly visits to the asylum, which is inspected by the Committee of Lunacy of the State Board of Public Charities, consisting of five members. (See p. 74.)

The original building, designed for 50 patients, cost 10,000 dollars,* or about £40 per bed, including the land. Considerable additions have been made, no doubt at a much higher figure. There are 84 acres attached to the institution, 40 of which are under cultivation.

The total number of patients is 95; 42 being men and 53 women. They are all of the private class. The total cost per week is £2 10s.

* The authority for this low figure is the Second Report of the Committee on Lunacy, Sept. 30, 1884.

The proportion of attendants to patients is 1 to 4. Their salaries: males, £3 4s. to £5 per month; females, £2 12s. to £3 14s. The female supervisor has £4 4s. a month.

With regard to restraint and seclusion, Dr. Hall is not an out-and-out non-restrainer, but resorts to restraint as little as he conceives possible. He said he did not find it necessary to restrain frequently, and that seclusion is not often resorted to. One woman was in seclusion, and she wore a camisole. She had been only a short time in restraint.

In reference to employment, it is stated that "few of the patients committed to this asylum are of the class that are accustomed to manual labour. In the opinion of the Superintendent, employment or occupation of the insane is of great importance, and he believes that great efforts should be made to secure it, and advocates the erection of a building adapted to the wants of a variety of patients, and securing opportunities for conducting the kind of employment best adapted to their general tastes and conditions."* Women are engaged in sewing and fancy work. There is a library of 800 vols., distributed in small sets of books throughout the wards. Patients are encouraged to take carriage drives.

The *Harrisburg Lunatic Hospital* is the State institution for the southern district of Pennsylvania, and was opened in 1851. Dr. Curwen was the medical superintendent for 30 years. The present superintendent, Dr. Gerhard, who was appointed in 1880, visited some English asylums several years ago.

Externally the building, which has a dome and handsome Tuscan portico, is attractive, but the same cannot be honestly said of the internal construction of this well-managed asylum. In fact, the best thing that could be done would be to raze the building to the ground, and erect several new ones in its place. I find from the Annual

---

* Second Report of the Committee on Lunacy of the Board of Public Charities of the State of Pennsylvania, Sept., 1884.

Report that an architect from Philadelphia has stated that the woodwork, in close contact with the steam pipes in the cellar, is a source of great danger from fire. He was also struck, it seems, with the "shallow, dark, ill-smelling, malaria-breeding vaults and passages," and " with the total absence in the wards of the proper appliances for successful ventilation, and the bad arrangements of the plan of the dormitories with reference to sunshine and pure air." Both he and another architect wisely recommend an entirely new building in place of the old. A Committee of the Senate of the House have concurred in this opinion. The cost is estimated at £100,000, or about £232 per bed. Dr. Gerhard, however, mentioned a much lower estimate.

From the dome there is a splendid view of the Alleghanies and the Susquehanna. The asylum is situated several miles from Harrisburg, on a plateau of ground seventy feet above the surrounding district.

There are four assistant medical officers; two in the male and two (women) in the female department.

The number of patients is 426: males, 210; females, 216. The building was originally designed for 300; accommodation for 200 more has been added.

The class of patients socially is mixed. The State requires that, in order of admission, the indigent insane shall have precedence of the rich. An Act of the Legislature (1883) limits the cost of the indigent insane to 16s. 8d. per week, which is stated to be sufficient. With clothing, the cost is 18s.; without, 17s. 1d.

		£.	s.	d.
4	private patients pay per week	0	10	5
84	,, ,, ,, ,, ,,	0	12	6
59	,, ,, ,, ,, ,,	0	14	7
23	,, ,, ,, ,, ,,	0	16	8
12	,, ,, ,, ,, ,,	1	0	10
1	,, ,, ,, ,, ,,	1	5	0

The proportion of attendants is one to ten. The salary of the male supervisor is £7 7s. per month, and of the female supervisor £4.

I found that, in regard to restraint and seclusion, Dr. Gerhard is a decided advocate of non-restraint save in very exceptional instances. On the male side three years have elapsed without his resorting to restraint, but it so happened that an exception to the general rule occurred in this department on the day of my visit. This was in a case of violent epileptic mania, in which Dr. Gerhard believed it best to restrain the patient's limbs. He was secured by straps to his bed. He was excessively abusive and excited. The complaints which he made did not refer to his restraint, but manifested an active antipathy to all around him. Hyoscyamine in doses of one-twentieth of a grain had been administered. I have since learnt that this attack passed away in a few hours, rendering restraint no longer necessary. In this asylum are no padded rooms, in one of which the patient would have probably been placed, in an English asylum.

On the female side of the house some form of restraint —camisole or wristlet—is not unfrequently in use. Three or four patients pass a very considerable part of their time in seclusion; one of these is an epileptic.

The recoveries have been 20·43 per cent.* since the opening of the institution. Dr. Gerhard is as sceptical as Dr. Earle in regard to the prevalent notions about the curability of insanity.

The deaths amount to 5 per cent., calculated on the average number resident.

Treatment.—With dirty patients, advantage is derived, as in some English asylums, from an enema administered on going to bed, followed by a suppository of five grains of tannic acid. It is not the custom to make this class get

---

* Two-thirds of the admissions last year were incurable, and half the remainder were doubtful cases. Only one-sixth were regarded as hopeful.

up during the night. In the Annual Report facilities are asked for giving different kinds of baths and for massage.

Employment.—There is a cobbler's shop, but only two patients were employed in it. About 62 men are employed in ward work; 45 join in games; 76 are engaged in reading and writing; one helps in the kitchen, and five in the laundry; 22 patients are employed on the farm and garden. Many of the women are engaged in needlework. Dr. Garver, one of the lady-assistants, desires to have a person qualified to teach calisthenics and dancing, and to play upon the piano. She observes in the last report that there is only one piano in the wards from which it is possible to extract a tune; "the other two are jangling and discordant abominations."

There is a service in the chapel every evening. The staff and about 80 patients were present when I was at the asylum, the medical superintendent officiating.

P.S.—Recent intelligence is to the effect that the old building is to be retained. If so, it is to be hoped some course may be decided upon to render the contingency of fire more improbable than it is under the present condition of the asylum, in referring to which the last Report of the "Committee on Lunacy" afresh lays stress on "its liability to become a crematory for the afflicted patients within its walls."

The asylum at Washington was visited on two occasions, and the pleasure derived from the sight of the beautiful buildings of the City was not marred by that of the *Government Hospital for the Insane*. It presents two prominent features. One is the architectural appearance and arrangements designed by the first superintendent, Dr. Nichols (now of Bloomingdale), who is thoroughly at home in establishments of this description, from the general outline down to the minutest detail. The main building is castellated in form, five storeys in height, in some portions four. On the windows are ornamental iron guards. A striking, and to an Englishman a pleasing, feature of the exterior, is the extent to which the walls are covered with ivy, an unusual circumstance in the United States.

Dr. Godding, formerly an assistant medical officer here, is the worthy successor of Dr. Nichols. He throws an ample stock of energy into the service of the institution, and is supported by able medical assistants, one of whom, Dr. Whitmer, visited our asylums in 1881. Ample opportunities were afforded me for obtaining information, while the guest of the superintendent for several days.

Another marked feature of the institution is the accommodation provided for different classes of patients in separate buildings, which merit a brief enumeration.

1. Detached Building for Idiots and women of colour (East Lodge).—Nine feeble children, black, and white.

The women of colour in this cottage are 15 in number; in the main building there are 80. Dr. Godding states that negroes are less excitable than whites.

The above is a brick house with stone facing. The basement is used for coals, &c. The house is heated from the steam boiler in the main boiler-house. A passage, under ground, communicates with the next building.

2. Atkins' Hall.—This was built in 1878 for the quiet working class of patients. It accommodates 61 patients, and was built at a cost of £1,500. This includes furniture, but is exclusive of land, it having been built on the estate. It is about 600ft. in length. Food will be taken to the East Lodge, that, namely, for idiots and coloured females, in an open cart through the tunnelled passage. There is a dormitory on the ground floor with 12 beds, which was intended for a day-room. Above is a much larger dormitory for 49. Dr. Godding states that this is the cheapest provision made for patients in the detached buildings, and that he could not duplicate it for the same sum, but could do it for £2000. The dimensions of the dormitory are 118ft. in length, 25 in breadth, and 11 in height.

3. "The Home," built for 150, or if necessary, 160 patients, at a cost of £13,000, including furniture, but not land. One half of the rooms are single. The rooms and corridors are very light. The water closets, &c., are detached. The building is fire-proof. It will be occupied in the course of a month or two.

It is proposed to have a separate building for criminal lunatics, and another building for convalescents.

Dr. Godding would like to have a separate dining-room in a distinct building, to which as many patients as possible would go from the other detached houses. This only depends upon a sufficient appropriation being made.

4. A separate building, called the West Lodge, the first erected separate from the main building, was built in 1856 for men of colour. It accommodates about 60.

5. The "Relief House."—This is built for the accommodation of 200 patients, at a cost of £10,000, or £50 a bed, including furniture, but not land. It was built in 1879. The excavations were made by patients. There are four storeys, three of which are occupied. There is a large dormitory with 27 beds, and another with 13. Each storey has back stairs. One patient in six has a single room. Partitions between rooms are of brick. There are 14 attendants, and a man and his wife are in immediate charge of the building.

In the fourth storey are work-rooms. Many brushes are made; also mats of husk or "shuck." There is also a billiard-room.

Summary of Numbers in Detached Buildings.

The East Lodge, for Idiots, &c.	24
Atkins' Hall	61
The Home	160
The West Lodge	60
The Relief House	200
	505

Thus, when the buildings are complete, the total accommodation will amount to about 500.

A detached building, called "The Rest," serves the purposes of a pathological room and mortuary. (See p. 81).

Dr. Godding, when speaking, in a recent report, of the detached buildings, observes " that some such provision for the quiet classes will be found a satisfactory solution of the great social problem of the care of the chronic insane. . . . What is wanted for them is care, and comfortable but inexpensive homes connected with our present curative establishments for the insane."

"The Relief Building, with outer and interior walls of brick, was constructed and furnished at a time when both labour and material were exceptionally low, and the whole expenditure did not exceed 250 dollars (£50) per patient. Allowing that this would be an unsafe figure on which to base an estimate at present prices, as it certainly would, still it is demonstrable that accommodations, which are all that are needed for the comfortable care of the mild cases of insanity, can be provided at an expense, for furnished buildings, of less than 500 dollars (£100) per patient. When it shall become the settled policy of the States to care for all their insane, taking them out of the town and county almshouses, placing them in homes connected with the existing hospitals, and giving them workshops and tillage lands, it will be a greater advance in their treatment than any that has been made since the earnest philanthropy of Miss Dix first called attention to the condition of this unfortunate class, and created so many of our present hospitals, which was a noble charity and meant for all. But practically it has been found that the liberal and expensive provision for the cure of insanity, admirable and necessary as it is for a part, has proved so great a tax that no State has thus far been willing to provide such elaborate asylums for the whole."

Government.—There is a Board of Managers consisting of nine members, including two ladies, appointed by the President of the United States. They meet twice a year.

Three of the Board inspect the asylum every month. There is no Lunacy Board.

There are 420 acres belonging to the hospital, of which the detached buildings and their land will eventually cover 12.

Number of Patients.—Males, 890; females, 260; total, 1,150.

Class of patients socially.—The asylum was originally designed for the army and navy insane. Now, 535 of the patients do not belong to this class. The District of Columbia contributes to the support of about 25 per cent. of the patients, the remaining 75 being paid for by the Government.

Class of Patients mentally.—Acute and chronic; about 25 per cent. are excitable.

Cost per week.—10s. 10d. for expenses per week, including ordinary repairs, but not patients' clothing.

Charge per week.—20s. to 50s. per week for paying patients.

Proportion of Attendants.—One to six or seven on the female side; one to nine throughout the asylum, exclusive of two supervisors and the night watches. In addition to the latter, Dr. Godding has devised the admirable plan of appointing a medical assistant to go through the male wards during the night. Being always dressed, he can at any time be called to see a patient on the other side of the house. In any difficult case an attendant is detailed to be in the bedroom during the night.

Restraint and Seclusion.—About ·5 per cent. (restraint). Six patients on an average are secluded for a part or the whole of each day.

Recoveries.—During 30 years 41·29 per cent. of the admissions. *Deaths.*—6·5 per cent. of average number of patients.

Many patients work on the grounds, others on the farm, and some are engaged in excavations. As regards brushmaking, there is at present a glut in the market, so that the patients have less to do. There is a tailor's shop; hair is prepared for mattresses; some patients paint. The work in the stables occupies some; others help in the laundry and kitchen. The women are employed in sewing, knitting, and fancy work. Thirty men on an average are employed in the different shops. Taking the male and female sides, it may be said that 25 per cent. are steadily employed. It is found that soldiers do not take kindly to work.

Amusements for the patients are steadily carried out; the room in which they are assembled is large and handsome. Thrice weekly during six months in the year there are dances, dramatic performances, and exhibitions of the magic lantern. On Sundays 500 patients attend the service held in the same room. Five ministers come in succession from Washington, and other denominations would be represented, were there any considerable number of patients belonging to them.

Before passing to the next State, I add the census returns of insane and idiotic in 1880 in the District of Columbia:—

	Insane.	Idiotic.
In Government Hospital for the Insane, Washington	860	13
Almshouses	4	4
Jails	0	0
Charitable Institutions	8	17
At Home	66	73
Total	938	107

The population in 1880 was 177,624.

A continuous journey of about twenty-four hours, through magnificent scenery, brought me from Washington to Chicago, where, through the kind forethought of Mr. Wines, facilities were afforded me for meeting Mr. McCagg, the President of the Board of Trustees of the *Eastern Hospital for the Insane at Kankakee*, Illinois, and

Dr. Dewey, the medical superintendent. With these gentlemen I proceeded to Kankakee, a town of 7,000 inhabitants (situated on a river of the same name), distant more than 50 miles from Chicago.

The central building, with its wings for both sexes, constructed in the linear form, is three-storeyed, and accommodates 275 patients. There are six wards in each wing, and, at the extreme end, three short wards. Each wing consists of two sections, but one of these sections, on the male side, has never been completed, and the capacity of the whole building is, in consequence, less by 75 beds than was intended. A covered corridor leads from the centre to the kitchen, bakery, &c., in the rear. From the kitchen there is a corridor to the engine and boiler-house. The laundry and bath-house are situated on the side nearest the detached buildings for the female patients, and the Recreation-Hall between them and the female wing; on the side looking to the male buildings is a carpenter's shop. The office of the clerk and storekeeper is further back.

The cost of the main and detached buildings about to be enumerated is stated by Dr. Dewey to have been as follows:—

(*a*.) Cost per bed, including all appropriations and expenditures of every kind, except running expenses to date (for 1,500 patients) . . £135

(*b*.) Cost per bed of all land and building of every kind (1,500 patients) £116

(*c*.) Cost per bed for buildings alone, including all appropriations made to date for erection of buildings of every kind (1,500 patients) . £112

(*d*.) Cost per bed of 18 detached buildings, accommodating 1,225 patients (including general dining-room and *employés'* quarters) . . £76

(*e*.) Cost per bed of all buildings of every kind to date, deducting north and south wings of main building (1,225 patients) . . . £103 (*f.*) Cost per bed of 31 institutions for the insane as given in report of Illinois Board of Public Charities for 1870. . . . . . £236

Dr. Dewey adds with regard to land that about £5,800 have been expended, the number of acres being 476.

It appears that Dr. Ray considered £200 to be the minimum cost per bed, in the States, for construction of buildings alone. It is maintained that no figures have hitherto shown a cost per bed more favourable than that at Kankakee, considering the substantial character of the dwellings. In reference to (*d*), it is remarked that the cost per bed (£76) of the 18 detached buildings for 1,225 patients is favourable, "when it is recollected that they are all substantial two-storey stone buildings (accommodating an average of 37 patients in 33 wards), with many brick partitions and all slate roofs, hard wood floors and basement concreted throughout, that they are connected with an ample system of sewage, have hot and cold water and gas in every part, and are provided with fire hydrants; also are arranged for thorough heating, either by steam or hot air furnaces."

I have already (pp. 80-1) mentioned the principal detached buildings at Kankakee. In all they number 18, and are built of stone, with slate roofs. Of these the following are devoted to the *female* patients:—

An Infirmary, providing 50 beds.

No. (1) Quiet and industrious, and some excitable; accommodation in four wards for 40 patients in each, a total of 160.

(2) Industrious and inoffensive; for 104 patients in two wards.

(3) Ditto, ditto.

(4) Quiet and convalescent; 44 patients; eleven single rooms (nearly all the bedrooms in the detached buildings are large dormitories with 10 to 50 beds).

(5) General dining-room for 500 patients, and kitchen; also rooms for 150 male and female *employés*.

Detached buildings for *male* patients are:—

An Infirmary, providing 50 beds.

A Relief building for 50 epileptic and 35 criminal patients; total, 85. These are in distinct divisions; and of the epileptics, the demented are separated from the more intelligent. The violent epileptics are in the main building. The ward for dangerous or criminal patients is built with special regard to security, and the windows are guarded. The epileptics sleep in one associated dormitory, and are placed under the care of a night watch. Adjoining this bedroom is the dormitory for the dangerous class. There are three attendants within call, should the night watch require assistance.

No. (1) provides 34 beds.
  (2)   ,,   31 ,, .
  (3)   ,,   46 ,, ; also a dining-room for 114.
  (4)   ,,   42 ,, ; ditto for 104.
  (5)   ,,   34 ,, ; more troublesome patients.
  (6)   ,,   31 ,, ; mostly employed in out-door work.
  (7)   ,,   44 ,, ; convalescent patients; eleven single rooms.
  (8)   ,,   160 ,, ; four wards, having 40 patients in each.
  (9)   ,,   104 ,, ; industrious and inoffensive (two wards).
  (10)   ,,   104 ,, ; ditto ditto.

To what has already been said as to the distribution of meals, may be added that in the detached buildings for males, the patients in the infirmary have their dining-room there; those in the relief building dine in that house, while the patients in Nos. 1, 3, and 5 use the dining-room for 114 in No. 3. The patients in Nos. 2, 4, and 6 use the dining-room for 104 in No. 4. Lastly, the patients in Nos. 7, 8, 9 and 10 dine in the " general dining-room " (p. 170).

In the detached buildings for females there is a dining-room in the infirmary for its 50 patients; there are two in No. 1, one of which is used by the inmates of this building, and the other for the patients in No. 2. Those in Nos. 3 and 4 dine in their own buildings, but their supplies come from the general dining-room kitchen, which also does duty for the other buildings in the same (the west) line, namely, 7, 8, 9, and 10 (males). The meat, however, is not cooked in this kitchen, but in the main central kitchen (more than 150 yards distant), which is used also for the dining-rooms of the several detached wards. The mode of conveying the meat in close metallic cars provided with a small coke stove, has been already described sufficiently at p. 80.

Then as to transit; in each dining-room hot tea and coffee are provided for on the spot, by an apparatus heated by steam or by gas. Provision will be made also for a warming closet in each dining-room.

The buildings are heated by steam, and hot water is carried to the detached buildings from steam generators.

The buildings are mostly about 85 feet apart. The patients who dine in the general dining-room would have to walk about 166 yards from the farthest, and 33 yards from the nearest, domicile they occupy.

The number of beds provided in the 18 detached buildings amounts to 765 for the male patients, and 462 for the females, a total of 1,227, the discrepancy in the provision for the two sexes arising from the non-completion of a section in the male wing of the main building. As the main building accommodates 275 patients, the total capacity of the whole institution is reckoned at 1,500; the number of inmates being 600.

It is maintained that the breaking-up of an establishment into moderate-sized detached buildings affords its insane population "a variety, a freedom, and a satisfaction not attainable in any hospital constructed upon the type now prevalent in the United States."

In the Report of the Board of State Commissioners of Public Charities of Illinois, issued in 1882, regret is expressed that the main building has been made so prominent. Were the work to be done over again, no centre building would be erected, and the wards designed for recent and curable, or refractory and troublesome, patients would be detached, and only two storeys in height. It is also regretted that the structures erected have not been even cheaper than is the case. At the same time the Kankakee buildings are sufficiently plain; and had less money been spent upon them, they would have been less durable and might have been exposed to the criticism made by Dr. Nichols in regard to some modern erections, of being "card-board shanties."

They are described correctly in one of the annual reports as being built in the style of an ordinary dwelling-house, two storeys high, with front and rear entrance and hall, verandahs, sitting-rooms on first floor, and sleeping-rooms upstairs. There are no guards on the windows, except in three rooms of the Relief building. The windows have an ordinary double-slung wooden sash, with panes of common glass, 16 x 20 inches.

The charge of the patients in the detached and main buildings is apportioned as equally as possible among the three assistant medical officers, one of whom, Dr. Bannister, supervises the detached cottages for men. Their salaries vary from £200 to £240 per annum, with rations. The Superintendent has £600 and rations. No retiring pension is given. The matron's salary is £120. The supervisor on the male side, who also acts as bailiff, receives £200.

Male attendants have from £4 to £6, and female from £3 to £4 a month. The proportion of attendants to patients is 1 to 10.

All patients, of whatever class, are free, the law of Illinois, as of some other States, providing care and treatment for the insane gratuitously. The cost per patient amounts to £40 per annum, including clothing and every other expense.

I found, on examining the record of restraint and seclusion, that during September, 1884, three male patients had worn mittens or muffs, or wristlets and belt, in each case chiefly for surgical reasons. This is about the average for the year. There had been no instances of restraint in July; in June three wore wristlets, and the same in May, the average time of restraint being eight hours. During April one man had been in restraint. No female patient had been restrained in August and September. In July there had been one instance of restraint for half-an-hour. The average amount of restraint during the whole previous year had been under 1 per cent. In regard to seclusion, there was during September an average of two patients a day for several hours on the female side, which was above the average for the year; while on the male side there was an average of under one a day.

It is stated in Dr. Dewey's report that restraint has been much abridged by constant efforts to employ destructive patients usefully, and by the introduction of strong dresses, locked clothing, &c., as in England.

In the detached buildings, the wards are in many instances open, and the number of patients on parole is very large, fully 40 per cent.

The ratio of patients employed in the detached buildings is stated to vary from 68 to 72 per cent., rising as high as 79 per cent.

The Trustees of the asylum, who are appointed by the Governor, meet monthly. The President, Mr. McCagg, although residing at Chicago, devotes a large amount of his time to the service of the institution. It is inspected by the State Board of Public Charities (Secretary, Mr. Wines) at least twice a year, and by a Committee of the State Legislature when in session.

This bald description of a very interesting experiment, conscientiously carried on by an excellent superintendent, is merely supplementary to the general remarks made upon the Cottage system, at p. 80, to which the reader is referred.*

The census returns of the insane and idiotic in Illinois (1880) are appended. At that time there were only 88 patients in the Kankakee Asylum.

	Insane.	Idiots.
In Hospitals for the Insane and Asylums	2195	31
Institutions for Idiots ...		306
Almshouses	749	411
Jails	12	21
Charitable Institutions ...	30	9
At Home	2148	3392
Total ...	5134	4170

The population of Illinois at the same date was 3,077,871.

From Illinois I proceeded to Wisconsin, the State of small county asylums—examples, as I have already said, of institutions under county care with State control. Having described them in a former chapter (p. 82), I do not propose to notice them further here, as their essential features have been sufficiently delineated. Whatever may be the ultimate verdict passed upon them, I regard it as a healthy sign when we see a number of laymen of the middle classes· taking an active personal interest in the insane, alive as I am to the dangers which arise when unsuitable cases are treated outside asylums possessing a medical head. One good feature of county asylums is that, generally speaking, the area of land on which the patients can work is considerable.

* To the reader interested in the Cottage system of providing for the insane, I would especially recommend Dr. Dewey's clearly and temperately written pamphlet, "Congregate and Segregate Buildings for the Insane," read at the Conference of Charities, held at Louisville, Ky., 1883. See also "The Alienist and Neurologist," 1884. A paper by Mr. Wines should likewise be read, "Provision for the Insane in the United States: A Historical Sketch," 1885.

There are two large institutions in Wisconsin, one a semi-State asylum, and the other a State asylum. To the former—the *Milwaukee Insane Asylum*, Wauwatosa—the following notes have reference:—

I visited this asylum, which is situated six miles from Milwaukee, in company with two members of the State Board of Charities, Prof. A. O. Wright, its energetic Secretary, and Mrs. Fairbank, of Milwaukee. The latter related to me, from her own personal experience, many striking instances of patients confined within a recent period in almshouses, frightfully neglected and treated like felons, who had been removed to asylums in consequence of the action of the Board, and rendered comparatively comfortable. In approaching the asylum itself, I saw the almshouse and a small home for 32 idiots close by. In the neighbourhood a private asylum is being built, of which Dr. McBride, the late superintendent of the asylum, will be the head.

This institution is a county asylum, but in consequence of having received assistance from the State, it may be called a semi-State asylum.

The assistant medical officer, Dr. Scriviner, was in temporary charge of the institution, and readily afforded all the information required. He has since been appointed superintendent.

The institution is governed by a Board of five Trustees, appointed by the Governor. The County Board of Supervision, consisting of 24 members, attends to the finances. Visitation is performed by the Trustees and by the State Board of Charities, consisting of five members, including one lady. There is no fixed time for their inspection.

The cost of building this asylum was £160 per bed, the capacity being 250, and the number of patients 300.

The building, of yellow brick, consists of three storeys, with stone basement, the ground attached amounting to 150 acres. It was opened in 1879. There are rooms in the centre of the building handsomely furnished for a few private patients. The provision made for the poorer patients appeared to be on the whole good.

The cost per week is 15s., and the charge for pay patients in the centre building is £2 to £5 a week; in the other wards it is £1 to £2 a week.

The proportion of attendants to patients is 1 to 10. There are too few in the refractory wards. The character of the attendants in this asylum appeared to me to be below the average. Their salaries range from £3 12s. to £5 a month for men, and £2 8s. to £3 for women. There are two night watches.

Several patients were in restraint, who, it seemed probable, might have had the free use of their limbs, had there been a sufficient number of good attendants. One man was seated in a strong room, fastened to his chair, his hands also being secured by a leathern muff. He was very tall, and was said to have proved himself a dangerous patient; he had torn up a portion of a floor, although, strange to say, one of his hands was totally disabled by an injury received before his admission. Another man, who was dancing about the gallery, had his hands secured by a leathern muff.

A number of statistical tables are given in the annual report, but as, unfortunately, they only refer to one year, it is useless to attempt to draw any inference from them in regard to either recoveries or deaths. It is extraordinary that the insertion of a table, comprising the movements of the inmates from the opening of the institution, or, at least, for a few years back, should be the exception, instead of the rule, in American reports.

The *Wisconsin State Hospital for the Insane* is situated at Mendota, on the lake of that name. It is a limestone building, surmounted by a small dome. There are six wards on each side. Their general appearance was home-like, and much has been done in the way of pictures and furniture. In some of the corridors there was a strong iron frame a few feet from the window at the end. There were pictures inside, and a table and chair, so that what is an undesirable construction, is thought in some instances to prove useful in separating one patient from others in the ward.

Dr. Buckmaster is the medical superintendent.

As this asylum is unfortunately crowded, some day-rooms have been converted into dormitories containing eight beds. In one corridor there were 51 patients with three attendants, and in another 56 patients with four. In the ward for excited cases there were 51 patients with four attendants.

In a ward in which were excited patients, there was the iron guard already described at one end of the corridor, but not at the other, thus leaving the glass window exposed, but the glass was not broken. On my remarking upon this, Dr. Buckmaster said the guard was retained only for the convenience of separation, and for placing plants and pictures within the enclosure. In the convalescent ward there were many plants, and ivy had been carefully trained over the window inside the corridor. There were ornamental guards outside the sashes, the panes of which were of large size.

There is one assistant medical officer.

The government is vested in a State Board of Supervision, which was appointed three years ago in place of a Local Board of Trustees then abolished. It has charge of the State institutions in Wisconsin. The members receive £400 a year, and all their expenses are paid. A member frequently resides in the asylum; in fact, one is almost constantly there. Visitation is made by the Governor of the State, who visits the asylum annually. The Legislature appoints a Committee before closing the session, whose duty it is to visit the hospital for the insane. They can appoint a physician to accompany and assist them. Further, there is the inspection by the State Board of Charities, the members of which are unpaid, with the exception of the secretary. The State Board of Supervision are Commissioners of Lunacy, and inspect monthly.

The farm consists of 400 acres. It may be mentioned that last year there were 250 tons of hay housed; upwards of 2,400 bushels of oats, 5,000 of mangold wurzels, and 1,200 bushels of potatoes. There are 75 acres devoted to Indian corn, which yielded 3,000 bushels. There are 80 cows—shorthorn, Jersey, and Alderney—300 pigs and nine horses. The garden covers 15 acres. Fruit is largely used.

The building cost about £200 per bed, exclusive of land. It was designed to accommodate 500 patients. Now there are 533; males, 289; females, 244. There is a great accumulation of chronic cases. It is stated that not more than 30 or 40 of the patients are curable. There are 12 general paralytics on the male side; none on the female. There are 40 epileptics.

The proportion of attendants to patients is 1 to 12. Their salaries range from £4 8s. a month to £6 for men, and about £3 to £4 4s. for women. The male supervisor, who has been at the asylum 14 years, receives £125 a year, while the female supervisor receives £62. One of the male attendants had been eight years at the Somerset County Asylum, England. I was informed that owing to the high wages obtained in other vocations it has been found difficult to procure sufficient help on the men's side.

Dr. Buckmaster considers that 10 or 12 would be a fair average of patients restrained.* There are generally three on the men's side. It should be stated that criminal lunatics are admitted into the asylum, and that as several are restrained by order of the Governor of the State, the superintendent has no choice. The forms of restraint are the camisole, the muff, and the wristlets. There are no less than 50 covered or crib-beds in use. The reasons assigned for restraint, as given in the carefully kept records

* More than 2 p.c., exclusive of crib-bed, the use of which has since been reduced to half the number.

of the institution, are to prevent destructiveness, the removal of dressings from wounds, self-injury, and escape. It is only just to the kindly superintendent to state that he has been a short time at the head of the asylum, and that he is proceeding to reduce the amount of restraint within much narrower limits, and this he will do, I am quite certain, with a right good will. In regard to seclusion, very little use indeed is made of it. Whatever may be the shortcomings of this asylum, or, rather, the sins of commission in the matter of restraint, there are no dark cells in which patients are immured and left to themselves, day after day, manacled and secluded at the same time.

The recoveries amount to 27·3 per cent. calculated on the admissions. There are many relapses. The deaths are at the rate of 6·12 per cent., reckoned on the average number resident. There is nothing to note specially in regard to treatment, but it may be stated that hyoscyamine, in doses of $\frac{1}{10}$ to $\frac{1}{20}$ of a grain, is occasionally administered. In one case alarming symptoms arose from the administration of $\frac{1}{45}$ gr.

With regard to employment, about 35 per cent. of men and women are employed during the greater part of the day, and about as many more partially, making 70 per cent. more or less occupied. The remainder go to the "Groves," and even those who are excited, and cannot or will not walk, are carried there. Quoits, swings, and croquet are among the outdoor amusements. Sometimes there is not a single patient in the house, except a few feeble invalids in bed. Excursions are made on Lake Mendota once or twice a week in summer; bathing and swimming in the lake are much enjoyed by the patients.

I understood Dr. Buckmaster to say that the religious services are usually conducted by the medical officers of the hospital. It would seem desirable that this should be always the case, for after a sermon on fasting (not by a medical officer) the tube had to be employed in nine cases for the purpose of forced alimentation!

Prior to the admission of a patient the proceedings are as follows:—Application is made on behalf of any person supposed to be insane, to the county court judge, &c., for a judicial inquiry as to his mental condition, or for an order of commitment to the hospital. It must be specified whether or not trial by jury is desired. The judge is to appoint two physicians to examine the alleged lunatic and report. If a jury is not demanded the judge may make an order of commitment. In the event of trial by jury, it must be in presence of the alleged lunatic and his counsel, the immediate friends, and medical witnesses, all other persons being excluded.

The census of 1880 shows that the number of insane and idiotic in this State amounted to:—

	Insane.	Idiotic.
In Hospitals and Asylums for the Insane	1230	30
Almshouses	315	126
Jails	22	2
At home	959	1627
Total	2526	1785

The population of Wisconsin was 1,315,497 in 1880.

The *Spring Grove Asylum*, Baltimore, is the State Hospital for the Insane for Maryland. It was re-organized in 1876. Dr. Gundry, formerly of the Dayton Hospital for the Insane, Ohio, has been superintendent for six years, and has done much to improve the appearance of the walls, and employed the patients in decorating them, as well as in other ways.

There are a considerable number of patients of colour in this asylum, the prejudice against whom is very strong—so strong that Dr. Gundry has been unable to break through it on the women's side of the house. These number 19, and have one attendant; their bedrooms are crowded. The number of men of colour is 11.

This institution, to which Dr. and Mrs. Gundry zealously devote their time and interest, would present a much more satisfactory aspect were it not so lamentably crowded. It is highly desirable that more money should be spent in furnishing it, and adding to the general appearance of comfort. Without a sufficient appropriation, the doctor's efforts are, to a large extent, paralysed.

The cost of the building and land (136 acres) was £450 per bed.

The Board of Managers, in their report for 1883, state that the small balance then on hand has necessitated " the leaving undone many matters of improvement and repairs which good order and proper preservation of the property called for, but which we could not accomplish owing to the limited appropriation by the last Legislature to the support of this great State charity." The crowded condition of the building is then referred to, and the absolute necessity of providing another hospital. It is stated that the limited grounds, absence of abundant water supply, and difficulty of sewage and drainage, forbid making increased accommodation at Spring Grove. It is added—and this is but too apparent—that "Maryland is largely behind many of the States in caring for the disabled poor within her borders." The Board urge that the State of Maryland should adopt the policy of caring directly for all its pauper insane. It appears that some of the counties are in arrear in their payment, and one of them was sued by the hospital several years ago. Dr. Gundry in his report makes the same complaints, and says that every year's delay only increases the evil.

Ventilation is not by any means so complete as in many American asylums; the fan is not in use. Earth closets only are resorted to, because the authorities do not allow the employment of drainage; and they work badly. Dr. Gundry proposes in his report that there should be a steam pump, with conducting pipes to various parts of the land, so that stations for distribution can be made at various distant points. With a view to extend the application of the sewage to the land, thorough underdraining will be necessary.

There are nine trustees or managers, one of whom, Mr. Francis White, kindly drove me to the institution, which is some miles from the city. The trustees, who are appointed by the Governor of the State, meet every month, and one of their number visits the asylum weekly. The grand jury pay a formal visit periodically. Mr. White regrets that there is no State Board of Charities.

There are 400 patients in the hospital—226 males and 174 females. Classification is according to mental condition. There is a mixed class of patients socially. The cost per head per week is 16s. The city and county pay £30 a year for each patient. There are 50 patients who pay from 24s. to £2 a week. The rest are maintained at the public expense.

The proportion of male attendants to patients is 1 to 14. There are two outdoor attendants. In the female department the proportion is 1 to 13; there is also a sewing girl.

The above includes the supervisor. Their salaries are:—Male supervisor, £5 8s.; ordinary male attendants, £4 8s. per month. There is no female supervisor. Female attendants have £2 12s. a month.

Dr. Gundry is strongly in favour of non-restraint. I was informed that there was little use made of seclusion, especially on the men's side. One patient is always secluded by express desire of the Board, he having once escaped and committed murder in the neighbourhood.

The recoveries are, unfortunately, not stated in the annual report for a longer period than the past year. The same may be said of the mortality.

About 80 patients are employed out of doors. Five men and eight women are employed in the laundry. Not quite 50 per cent. of the total number of patients are more or less employed. Dramatic scenes are represented in the recreation-room, which is used also for the chapel. The magic lantern is a favourite amusement.

The mode of proceeding in Maryland in regard to the admission of paupers into asylums is as follows:—" It is enacted that when any person is alleged to be a lunatic or insane pauper, the Circuit Court for the county in which said person may reside, or the Criminal Court of Baltimore, if he be resident in the city, shall cause a jury to enquire whether he is insane, and if found so shall cause him to be sent to the almshouse or hospital, &c., until he have recovered. Nothing in this provision shall prevent the friends of such lunatic from confining him or providing for his comfort."*

It is said that the almshouses in Maryland are in a very bad condition. There is one at Baltimore. A gentleman informed me that when he visited it a year ago the insane inmates were in an unsatisfactory condition, and that some were chained. A separate receptacle for the insane paupers is in course of erection near the almshouse.

Another institution for the insane at Baltimore is the *Mount Hope Asylum*, an institution under the charge of Roman Catholic Sisters of Charity. The building, situated a few miles from Baltimore, is a very extensive edifice of red brick, five storeys high. It stands in 400 acres of land, and it provides accommodation for 546 patients, the females preponderating.

One feature of the asylum noticeable on passing through the first storey, a ward for 31 patients of the worst class, is that they are attended to by not only a male attendant, but by two Sisters. This plan of having female attendants in the male department is general throughout the house, and is regarded as an important and successful arrangement. On the female side of the house there are two Sisters and one help in each ward. Pictures were on the walls, and the floors were carpeted.

* See Harrison's Collection of Lunacy Laws, 1884.

The rooms as well as the corridors were very tastefully furnished. The windows of the strong rooms were high; none of these rooms were without a bed. The rooms in the storey above, for a rather better and more curable class, were cheerful. The dress of the patients was very neat; one man was playing the violin surrounded by a group of patients. The windows are protected by ornamental iron frames throughout. In the rooms in the roof were 31 patients who were cared for by three Sisters and one male attendant. The day-room here also was carpeted. In some single rooms distinct from the ward there were two patients in seclusion, but not otherwise restrained. In another gallery containing about the same number of patients one was playing the violin to the rest. The beds were clean and comfortable. There was a large dormitory in the attic, which was somewhat bare. The view from the roof, on which there is a cupola, commands an extensive prospect of the surrounding country.

The number of wards is sixteen. Patients are classed according to mental condition.

There is one assistant medical officer. Dr. Stokes, of Baltimore, is the visiting physician, and has been so about forty years. He attends three times a week, and the assistant medical officer, who resides near the asylum, visits on the other days of the week.

The weekly charge for paying patients, of whom there are 290, varies from 20s. to £3 a week. The city cases (156) pay 12s. 6d. a week which includes clothing, while 100 of the patients pay next to nothing.

No patient was in restraint. Several were in seclusion.

Recoveries.—Here, as so generally is the case, the annual report affords no information as to the number of patients recovered during a term of years. During the past year 65 patients were discharged recovered; 53 of these being recent, and 12 chronic cases. The admissions during the year were 184. It is stated by Dr. Stokes that he has often proved in his reports that 70 or 80 per cent. of acute cases admitted within three months after the inception of the disease have recovered, and that, when the case was not complicated with organic disease of the brain, even 90 per cent. have been cured. We are not informed whether this high percentage is calculated on single admissions; in other words, whether it has reference to cases or persons. The statement is, indeed, made that it seldom happens that patients are re-admitted on a recurrence of the disorder. This experience is so contrary to that of every other asylum I know to which patients, if they do relapse, would return, that I should have greatly valued statistical tables covering the history of the asylum from its commencement to the present time.

As regards the mortality, also, we have only the deaths for the past year, namely, 45; but as the average number resident is not given, we are quite precluded from obtaining a correct percentage for even one year. We can only tell that it was 7·19 per cent. of the whole number under treatment.

Dr. Stokes is fully alive to the importance of labour on the land as an aid to recovery. He observes that the farm, the garden, the laundry, the kitchen, and the sewing-room furnish the means of occupation, by which the recovery of convalescence is advanced, and the incurable are relieved from the misery attendant on a state of idleness. No figures, however, are given to show the actual number of patients employed.

There is a Recreation Room distinct from the chapel. It constitutes a large central parlour. Once a week evening-entertainments are given. Billiards on the male side, and pianos in the female department, are provided, and appreciated by the patients.

The so-called "Picnic Grounds," situated in a retired situation not far from the asylum, are stated to be very popular. Tennis and croquet are provided, also boats for rowing on the lake, which, although "*manned*," are each under the care of the Sisters, and are much used in summer. There is also a pavilion for dancing, and a cottage-kitchen in which meals are prepared.

The form of medical certificate is extremely simple, namely, "I hereby certify that I have seen and examined A.B., of C.D., and believe him to be insane and that he ought to be placed for treatment in a hospital for the insane."

The order for admission, to be signed by a citizen of Baltimore, is as follows: "I request that the said A.B. shall be admitted into the 'Mount Hope Retreat,' and in consideration of his being so admitted, I hereby guarantee the performance of the above obligation " (a bond to pay for board, &c.).

It only remains to add the statistics of insanity and idiocy in Maryland as given in the census of 1880:—

	Insane.	Idiots.
In Hospitals for Insane and Asylums	912	37
Almshouses	200	86
Jails	9	0
Charitable Institutions	5	9
At Home	731	1187
	1857	1319

The population of this State at the above date was 934,943.

As this is the last State the figures of which I give in relation to the provision for the insane, I would add that, throwing together the figures of all the States visited, the result shows that there were at the last census 44,824 insane. Of these there were in hospitals for the insane and asylums 22,798, and in almshouses 5,531. There were 57 in jails, 90 in charitable institutions, while 16,348 were at home. In the same States were 25,605 idiots, of whom 421 were in hospitals for the insane and asylums, 2,406 in almshouses, 24 in jails, 124 in charitable institutions, 1,514 in institutions for idiots, and 21,116 at home.

The figures, although they accord in their general character with those obtained from the aggregate census for the whole of the States, show some difference in a favourable direction, seeing that more than one-half were under care in hospitals or asylums for the insane. I am surprised, however, that in the States which I visited, the proportion of the insane in almshouses is as high—namely, 10 per cent.—as in the total number of States. The number at home or in private families was 32 instead of 44 per cent.; favourable by comparison, but still a large proportion.

The total population of the States I visited was, in 1880, 19,364,206, and as the number of insane and idiotic in these States amounted to 70,429, there was 1 person of unsound mind to 274 of the general community—a somewhat higher proportion than that in the total States. (Appendix C.)

In England and Wales the proportion of insane in hospitals and asylums (including idiots) in 1884 was 70 per cent. The number in ordinary workhouses was 15 per cent., or 5 per cent. higher than in the States. I think it may be said that our workhouses are, as a rule, fit receptacles for the class of patients sent there.

For more detailed information in regard to the insane and idiotic in the United States, the reader is referred to the Appendix.

CHAPTER V.

THE INSANE IN CANADA.

Province of Quebec.

On the 30th August last (1884) I visited the lunatic asylum at Longue Pointe, seven miles from Montreal, called the *Hospice des Alienés de St. Jean de Dieu*. It was built by the Sœurs de Providence, and opened in 1876. The Province of Quebec contracts with them to maintain the lunatic poor* in one of the two districts into which the Province is divided; the asylum at Beauport, near Quebec, providing similarly for the other district. Private patients are admitted. The building—which, surmounted by three cupolas, is a prominent object from the St. Lawrence in approaching Montreal from Quebec—is built of red brick, and consists of a centre and wings. Some of the latter were added three or four years ago; others are now in course of erection, and will not be finished till the end of the year. Dr. Henry Howard, the visiting physician, kindly facilitated my desire to see the asylum, and escorted a small party, consisting of Dr. Ross of Montreal, Dr. S. Mackenzie of London, and myself, to the institution. I must express to Dr. Howard my lasting obligations for his attention and assistance. We were received by the Lady Superior, Sœur Thérèse de Jesus, who had been apprised of our visit. She conducted us

---

* At the rate of 100 dollars or £20 per annum per head at Montreal and 130 dollars at Quebec—a very insufficient sum, it would seem, for board, lodging and clothing. I understand that the money originally borrowed of the Provincial Government by the Montreal Asylum has been refunded, and that money has been borrowed from private quarters to assist in the erection of the additional buildings.

through the building, and was most courteous in her manner, and in replying to the numerous questions with which I troubled her. I am glad to have this further opportunity of thanking her and the other Sisters for their kindness throughout the visit.

The neatness and cleanliness of the hall, reception-room, and office strike the visitor very favourably on entering the establishment. The *Apothecaire* is a model in these respects. The Sisters have themselves published a pharmaceutical and medical work, a large volume entitled "Traité Elémentaire de Matière Médicale et Guide Pratique," a copy of which the worthy Lady Superior was good enough to present to me. I was somewhat disappointed to find, on examining its pages, that only one was devoted to mental alienation, of which nine lines suffice for the treatment of the disorder. Among the moral remedies, I regret to see that *"punitions"* are enumerated; their nature is not specified. Two skeletons in the *Apothecaire* were shown to us by Sœur Thérèse, as being much valued subjects of anatomical study for the nuns, who would, it is not unlikely, consider their knowledge of the medical art sufficient for the needs of the patients. The law, however, obliges a medical man to reside in or near the asylum. Dr. Perrault, whom we did not see, occupies this post. This officer is appointed and paid by the Sisters; the visiting physician, on the contrary, is appointed and paid by the Provincial Government. We looked down upon a very large kitchen, where cooking by steam was going on actively, and a favourable impression as to the supplies was left upon the mind by the busy scene which presented itself. The amount of vegetables (potatoes, turnips, cabbages, &c.) produced on the land, is very large—more potatoes, I believe, than they consume. Maize, wheat, oats and buckwheat are raised. The estate consists of 600 acres. There is a large number of cows, and the asylum buys beasts to fatten and kill, thereby saving a considerable sum. I was informed that about fifty patients were usually employed out of doors, and more in harvest time. That such an establishment should be conducted by nuns must seem remarkable to those who are unacquainted with the large part taken by Sisters of Charity in the management of hospitals in countries where the influence of the Roman Catholic Church extends. Theoretically, it would seem to be an admirable system, and to afford, in this way, a wide field for the employment of women in occupations congenial to their nature, and calculated to confer great advantages upon the sick, whether in mind or body. That women have an important *rôle* in this field will not be denied; but experience proves only too surely that to entrust those of a religious order with administrative power is a practical mistake, and leads to abuses which ultimately necessitate the intervention of the civil power.

The asylum consists of a succession of corridors and rooms similarly arranged, there being dining-rooms, recesses, and single and associated dormitories. There are four storeys uniform in construction, exclusive of the basement and the rooms in the roof, and these four are supplied with open outer galleries or verandahs, protected by palisades. The lower storeys are clean and well furnished, and the patients appeared to be comfortable. The apartments of the private patients were, of course, the best furnished. It was curious to see in the day rooms on the male side a nun with a female assistant. They are in the wards all day, and sleep together in another part of the building. In the refractory ward for men there were two male attendants, and in the other wards one male attendant, in addition to the two females. In each ward on the women's side there were two assistants with the nun in charge, and in the refractory gallery there were three assistants. The nuns and female assistants are not paid. The corridors, the width of which was fair, were carpeted down the centre, and there were pictures on the walls in considerable number. In the day-rooms, on the floors of which was oilcloth, the furniture, though simple, was by no means insufficient. In the recesses of the corridors, as well as in the corridors themselves, were seats for the patients. Although there were rooms on both sides of the corridor, the latter was fairly lighted by the recesses, &c. The dormitories were very clean, and presented a neat appearance; the beds were of hair, and a bright-coloured counterpane had a pleasing effect. Single rooms, used as bed and sitting-room, were very neatly furnished, and had every appearance of comfort. For paying patients, and for a considerable number of the poorer class, I have no doubt the accommodation is good, and as I must shortly speak in terms of strong reprobation, I have pleasure in testifying to the order, cleanliness, and neatness of those parts of the building to which I now refer, and over which we went in the first instance.

It is as we ascend the building that the character of the accommodation changes for the worse. The higher the ward, the more unmanageable is the patient supposed to be, and the galleries and rooms become more and more crowded, and look bare and comfortless. The patients were for the most part sitting listlessly on forms by the wall of the corridor, while others were pacing the outside gallery, which must afford an acceptable escape from the dull monotony of the corridor. The outlook is upon similar galleries in the quadrangle at the back of the building; and, to a visitor, the sight of four tiers of palisaded verandahs, with a number of patients walking up and down the enclosed spaces, has a strange effect. These outside galleries are, indeed, the airing courts of the asylum. There are no others. If the patients are allowed to descend, and to go out on the estate, they usually do so in regular order for a stated time, in charge of attendants, like a procession of charity school children. Those who work on the farm must be the happiest in the establishment.

In the fourth tier were placed the idiots and imbeciles —a melancholy sight necessarily, even when cared for and trained in the best possible manner, but especially so when there is no attempt made, so far as I could learn, to raise them to a higher level or educate them. If, however, they are kindly treated and kept clean, my regret on account of educational neglect would be much less than the pain caused by the state of the patients and their accommodation in the parts of the establishment next described. Far be it from me to attribute to these Sisters of Charity any intentional unkindness or conscious neglect. I am willing to assume that they are actuated by good motives in undertaking the charge of the insane, that they are acute and intelligent, and that their administrative powers are highly respectable. Their farming capacities are, I have no doubt, very creditable to them. It is not this form of farming to which I have any objection or criticism to offer. In the vegetable kingdom I would allow them undisputed sway. It is the farming-out of *human* beings by the Province to these or any other proprietors, against which I venture to protest.

It is impossible to convey an adequate idea of the condition of the patients confined in the gallery in the roof, and in the basement of this asylum. They constitute the refractory class—acute and chronic maniacs. They and the accommodation which has hitherto been provided for them, must be seen to be fully realized. To anyone accustomed to a well-ordered institution for the insane, the spectacle is one of the most painful character. In the course of seven-and-thirty years I have visited a large number of asylums in Europe, but I have rarely, if ever, seen anything more depressing than the condition of the patients in those portions of the asylum at Longue Pointe to which I now refer. I saw in the highest storey, that in the roof, a gloomy corridor, in which at least forty refractory men were crowded together; some were walking about, but most were sitting on benches against the

o

wall or in restraint-chairs fixed to the floor, the occupants being secured to them by straps. Of these seated on the benches or pacing the gallery, a considerable number were restrained by handcuffs attached to a belt, some of the cuffs being the ordinary iron ones used for prisoners, the others being leather. Restraint, I should say in passing, was not confined to the so-called refractory wards; for instance, in a lower and quieter ward, a man was tightly secured by a strait-waistcoat. Dr. Howard had him released, and he did not evince any indication of violence. It was said he would tear his clothes—a serious matter in an asylum conducted on the contract system! The walls and floor of the corridor in the roof were absolutely bare. But if the condition of the corridor and the patients presented a melancholy sight, what can be said of the adjoining cells in which they sleep and are secluded by day? These are situated between the corridor and a narrow passage lighted by windows in the roof. Over each door is an opening the same width as the top of the door, and three to four inches in height, which can be closed or not as the attendant wishes. This aperture is, when open, *the only means* of lighting the cell. The door is secured by a bolt above and below, and by a padlock in the middle. In the door itself is a *guichet* or wicket, secured, when closed, by a button. When opened, a patient is just able to protrude the head. There is, as I have intimated, no window in the room, so that when the aperture over the door is closed it is absolutely dark. For ventilation, there is an opening in the wall opposite the door, which communicates above with the cupola; but whatever the communication may be with the outer air, the ventilation must be very imperfect. Indeed, I understood that the ventilation only comes into operation when the heating-apparatus is in action. What the condition of these cells must be in hot weather, and after being occupied all night, and, in some instances, day and night, may be easily conceived. When the bolts of the door of the first cell which I saw opened were drawn back and the padlock removed, a man was seen crouching on a straw mattress rolled up in the corner of the room, a loose cloth at his feet, and he stark naked, rigorously restrained by handcuffs and belt. On being spoken to he rose up, dazzled with the light; he looked pale and thin. The reason assigned for his seclusion and his manacles was the usual one, namely, " he would tear his clothes if free." The door being closed upon this unfortunate man, we heard sounds proceeding from neighbouring cells, and saw some of their occupants. One, who was deaf and dumb, as well as insane, and who is designated *l'homme inconnu*, was similarly manacled. In his cell there was nothing whatever for him to lie or sit upon but the bare floor. He was clothed. Some of the cells in this gallery were supplied with bedsteads, there being just room to stand between the wall and the bed. When there is no bedstead a loose palliasse is laid on the floor, which may be quite proper. In reply to my inquiry, the Lady Superior informed me that it was frequently necessary to strap the patients down in their beds at night.

Passing from this gallery, which I can only regard as a "chamber of horrors," we proceeded to the corresponding portion of the building on the female side. This was to me even more painful, for when, after seeing the women who were crowded together in the gallery, on benches and in fixed chairs, many also being restrained by various mechanical appliances, we went into the narrow passage between the pens and the outer wall, the frantic yells of the patients and the banging against the doors constituted a veritable pandemonium. The effect was heightened when the *guichets* in the doors were unbuttoned, and the heads of the inmates were protruded in a row, like so many beasts, as far as they could reach. Into this human menagerie, what ray of hope can ever enter? In one of the wards of the asylum I observed on the walls a card, on which were inscribed words to the

effect that in Divine Providence alone were men to place
their hopes. The words seemed to me like a cruel irony.
I should, indeed, regard the Angel of Death as the most
merciful visitant these wretched beings could possibly
welcome. The bolts and padlocks were removed in a
few instances, and some of the women were seen to be
confined by leathern muffs, solitary confinement not being
sufficient. One of the best arguments in favour of restraint by camisole or muff is that the patient can walk
about and need not be shut up in a room, but we see here,
as is too often seen, that unnecessary mechanical restraint
does not prevent recourse being had to seclusion. A cell,
darkness (partial or total), a stifling atmosphere, utter
absence of any humanizing influence, absolute want of
treatment, are frequently the attendants upon camisoles, instead of being dispensed with by their employment. When such a condition of things as that now
described is witnessed, one cannot help appreciating,
more than one has ever done before, the blessed reform in
the treatment of the insane which was commenced in
England and France in 1792, and the subsequent labours
of Hill, Charlesworth, and Conolly. But it is amazing to
reflect that although the superiority of the humane mode
of treating the insane, inaugurated nearly a century ago,
has been again and again demonstrated, and has been
widely adopted throughout the civilized world, a colony
of England, so remarkable for its progress and intelligence as Canada, can present such a spectacle as that I
have so inadequately described as existing, in the year of
grace 1884, in the Montreal Asylum.

Before leaving the asylum, I visited the basement, and
found some seventy men and as many women in dark,
low rooms. Their condition was very similar to that
already described as existing in the topmost ward. A
good many were restrained in one way or another, for
what reason it was difficult to understand. Many were
weak-minded, as well as supposed to be excitable. The

patients sat on benches by the wall, the rooms being bare and dismal. A large number of beds were crowded together in a part of the basement contiguous to the room in which the patients were congregated, while there were single cells or pens in which patients were secluded, to whom I spoke through the door. The herding together of these patients is pitiful to behold, and the condition of this nether region must in the night be bad in the extreme. I need not describe the separate rooms, as they are similar to those in the roof. The amount of restraint and seclusion resorted to is, of course, large. Yet I was informed that it was very much less than formerly.

To the statement in regard to the crowding of the patients in this asylum, it will be objected that I have given a description of a state of things which will shortly disappear, as additional wards are being provided for their accommodation. While I am glad to hear that other rooms will be available before long, I am not by any means convinced that the lowest and topmost wards of this asylum will be disused for patients. There are now, the Lady Superior said, about 1,000 lunatics in the buildings; and when first informed that new wings were being prepared, I concluded that it was for the purpose of providing increased accommodation for the existing number of inmates only. That hope, however, was greatly lessened, if not wholly dispelled, when I learnt from this lady that when these new wards are ready there will be room in the institution for 1,400 patients. It is said the new rooms will contain 600 beds, but how many cubic feet are allowed in this calculation I do not know. I have no hesitation in saying that when the patients are removed who now occupy the two portions of the building I have described, and when the occupants of the other galleries are reduced to the number the latter ought properly to accommodate, there would be at least 400 patients who should be removed from the old to the new building. If I am correct in this opinion, the present lamentable evils will continue after the opening of the additional apartments, or if they are mitigated for a time they will but too surely be renewed as fresh admissions take place. Assuming, however, that overcrowding is lessened, and that these dark cells should cease to be used, what guarantee—what probability—is there that the manacles will fall from the wrists of the patients of this asylum? I am not now speaking from the standpoint of absolute non-restraint in every conceivable instance of destructive mania. It is sufficient to hold that the necessity for mechanical restraint is exceptional, and that in proportion as an asylum is really well managed the number whose movements are confined by muffs, strait-waistcoats, and handcuffs will become smaller and smaller. The old system of treating the insane like felons has been so completely discarded by enlightened physicians devoted to the treatment of the insane, that it can no longer be regarded as permissible in a civilized country. The astonishment which I experienced in witnessing this relic of barbarism in the Province of Quebec, is greatly increased when I see such excellent institutions as the lunatic asylums of the adjoining Province of Ontario. I am perfectly certain that, if it were possible to transfer the worst patients now in the asylum at Montreal to these institutions, they would be freed from their galling fetters and restraint-chairs. They would quit their cells also, and, in many instances, be usefully occupied where they are now restrained, with the result that in not a few cases perfect recovery to health would follow. "Look on this picture and on this," were words constantly in my mind after visiting the institutions of the two Provinces. It can hardly be urged that a system which is attended by great success in the one Province, would be less successful in the other.

The recoveries at the Montreal Asylum during the last ten years have been decidedly few; while the mortality has been above the average of asylums, namely, 11 per cent. of the average number resident.

The question arises, why this difference in the condition of the insane in the asylums of the two Provinces? Whatever other reasons there may be for this extraordinary contrast, I have no doubt that the main cause is to be found in the different systems upon which the financial management of these institutions is based. It is a radical defect—a fundamental mistake—for the Province to contract with private parties or Sisters of Charity for the maintenance of lunatics. This, it cannot be too often repeated, is the essential root of the evil; and unless it be removed, the evil, although it may be mitigated, will remain and will bear bitter fruit. If any steps are to be taken to remove the present deplorable condition of the insane in the asylum of Montreal, it must be by the Province taking the actual responsibility of these institutions into its own hands. Whatever may be the provision made by private enterprise for patients whose friends can afford to pay handsomely for them, those who are poor ought to have the buildings as well as the maintenance provided for them by the Legislature. They are its wards, and the buildings in which they are placed should belong, not to private persons, but to the public authorities, with whom should rest the appointment of a resident medical officer.

The official inspection of this institution must now be referred to. There are, I am informed, three inspectors of the asylums and prisons of the Province, namely, Drs. L. L. L. Desaulniers, A. de Martigny, and Mr. Walton Smith. They report to the Provincial Secretary, who resides at Quebec, and is the Government officer to whose department these institutions pertain.* I was informed

---

* I little knew at the time of my visit that a correspondence had recently taken place between the Lady Superior, the Provincial Secretary, and Dr. Howard in regard to the improper detention of patients in the asylum. This correspondence has since appeared as an official document, and reveals in a startling manner the character of the system pursued at this institution. Dr. Howard, it seems, represented to the Lady Superior that in

that the visits of the inspectors are due three times in the year. The Grand Jury are empowered, when they meet, to visit asylums and make a presentment to the Court in regard to their condition, but I understand that this is generally a very formal proceeding. With regard to the authority of the visiting physician appointed and paid by the Government, it has been hitherto, so far as I could ascertain, almost, if not entirely, *nil*. His hands have been so tied that he could not be held responsible for the way in which the asylum has been managed. The Quebec Legislature passed an Act in June, 1884, which has only just come into force, and which, among other provisions, extends and enforces the authority of this officer. It remains to be seen whether this Act invests him with sufficient power to carry out any system of treatment or classification of the patients which he may deem requisite.

There should, however, in any case, be a medical superintendent, with competent knowledge of the treatment, moral and medical, of the insane, with undivided authority and responsibility inside the institution, although subject to the Government, aided by efficient medical inspection.

Should the contract system be abolished, should capable medical men be placed at the head of the institutions of

---

his opinion a considerable number of patients had recovered, and should be discharged, while others should be liberated who had never been insane at all. This recommendation was met by refusal. Appeal was then made to the Provincial Secretary, who supported the physician. For some time Sister Thérèse refused compliance with the demands of even the Provincial Secretary. On one occasion vehicles were sent to the asylum to convey those patients away for whose removal instructions had been given, but the heart of the fair Pharaoh of Longue Pointe was hardened, and she refused to let the people go. Only by the stringent exercise of civil authority was the saintly Sister at last reduced to obedience, protesting meanwhile that the consequences to society, of letting loose some thirty or forty patients, would be disastrous. The system under which such a state of things can exist stands self-condemned, however willing one may be to accede to the truth of the statement of the Chaplain of the asylum, the Rev. F. X. Leclerc, when he writes in a recent Report: "Everyone knows that the object to which the lives of these worthy Sisters are devoted is to personify, with the unfortunates, that Divine Providence whose glorious name they bear."

the Quebec Province, and should inspection made by efficient men be sufficiently frequent and searching, the asylums for the insane of this Province will become institutions of which Canadians may be justly proud, instead of institutions of which they are now, with good reason, heartily ashamed.

Beauport Asylum, Quebec.

I visited the Beauport Asylum, at Quebec, August 18th, 1884. It was established in 1845; additions were made to the original building, in 1865 for the male patients, and in 1875 for the female patients. The medical superintendents reside in the city, several miles away, and I had not the pleasure of seeing them. There are two visiting physicians. The asylum is inspected by Dr. Desaulniers, Dr. A. de Martigny, and Mr. Walton Smith. Resident on the premises is the warden, and in the vicinity is an assistant physician. I have to express to both these gentlemen my obligations for the kind way in which they received me, the time they devoted to my visit, and their readiness to show me the various parts of the building. My thanks are especially due to Mr. A. Thomson of Quebec, for the assistance he rendered and for accompanying me.

The asylum is a striking object to visitors to the Montmorency Falls, as they pass along the road where it is situated. The approach is pleasant and the entrance attractive, being marked by the taste and cleanliness which characterise the dwellings of the Canadians generally. The warden received us politely, and took us round the building devoted to female patients. His wife occupies the post of matron, and has two assistants under her. The corridors into which we first went are sufficiently spacious, and serve the purpose of day-rooms to a large extent, the patients being seated or walking about. The patients here were well dressed, and appeared to be as comfortable as their condition would allow. The associated dormitories are large, cheerful rooms, well ventilated, and the beds neat and clean. I supposed that the linen had been clean that morning, but was informed that it was the last day in use, and that it was changed weekly. Strips of carpet and mats in the dormitories, as well as in the corridors, relieved the bareness of the floor.

The position and construction of a series of single bedrooms attached to the wards, are most unfortunate. They are placed back to back, so that there is no window in them, the narrow passage which skirts them receiving light from a window at either end. There is an aperture over the door, and a small one in the door itself. The ventilation is most imperfect, and it was not denied that in the morning their condition is the reverse of sweet. Some of these cells—for cells they must be called—were very close when I visited them. How such rooms came to be built for lunatic patients, for whom good air and sufficient light are so important, it is difficult to comprehend. I was informed they were planned to expedite the escape of the inmates in case of fire, there having been a conflagration some years ago in which twenty-six patients perished, but I failed to see the fitness of such an arrangement. It appeared to me to be due to the desire to economise room, and I am not surprised to find, from one of the annual reports given me, that credit is claimed on the ground that the cost for care and maintenance is less than at ten asylums with which the Beauport Asylum is compared.

I have spoken favourably of the associated dormitories opening into the corridors. Those, however, in the attic were very gloomy and crowded with beds. I have also referred favourably to the dress of certain patients. I must add that in some parts of the house they were barely clad, and presented a very neglected appearance altogether.

The number of women in restraint was very considerable. Some wore the *manchon* or muff, others the close glove (*mitaine*); others were restrained by leather wristbands (*poignet*) fastened to a belt, while some were secured by the *gilet de force*, so that movements of the arms were effectually prevented. Several were secured to the bench on which they were seated. In one small airing court upon which I looked down, not a few were restrained; the whole company appeared to be unattended, or if there were attendants, the latter did not consider it a part of their duty to keep their dress in decent order. In referring to mechanical restraint, I beg it to be distinctly understood that neither here nor in any other asylum do I judge of the condition of the patients from the total non-restraint point of view. The amount resorted to in this asylum would not be seriously justified by any physician of the insane, with whom I am acquainted, whatever his views on non-restraint may be.

After leaving the building in which the women are located, we walked into the grounds over a stream to a steep, grassy, airing court, which was fortunately shaded from the blazing sun of that day. Here a number of female patients were congregated, with one or two attendants. A wooden fence separates this ground from a corresponding grass plot for the men. From a window in the building for the latter sex I looked down subsequently, and the sight of the female patients lying or sitting on the grass in unseemly attitudes, and with scant and neglected attire, did not commend itself as one altogether desirable. The number of attendants is quite insufficient, and I cannot say I was favourably impressed with their appearance. Where so much importance is attached to economy, this cannot excite surprise. Their pay is very insufficient, as well as their number.

Passing to the building for the male patients, over which the resident physician escorted us and manifested the greatest willingness to show every corridor and room, I would observe that there are certain wards which, like those for the corresponding class of women, are both clean and respectably furnished; but when I have done justice to the accommodation afforded in these galleries, I have said all that I can say in the way of commendation. The higher one ascended in the building, the lower the condition of the patient—the corridors were much crowded, and the amount of mechanical restraint was excessive. In the worst ward, the sight was in the last degree painful to witness. Here were some thirty patients. Some had leathern muffs, others the belt and *poignet*, while several were in cells as dark as those on the women's side, and were also restrained. One had his legs fettered at the ankles. There were also several men in restraint-chairs, to which they were fastened, and not only so, but they wore muffs. They were in their shirts, and over their exposed persons flies were crawling in abundance—a spectacle which it must suffice to mention without characterizing further. Among patients of the class now referred to, I counted fourteen restrained, but I do not pretend to have noted them all. For a man who was given to scratching his face, it was considered necessary not only to secure his hands by the muff, but to place him in a crib-bed.

But it is needless to describe in more detail an institution which, however willingly I may praise where praise is due, is so radically defective in structure and so fundamentally different from any well-conducted institution of the present day, in the matter of moral, to say nothing of medical treatment, that no tinkering of the present system will ever meet the requirements of humanity and science. I regret to write thus. It is a thankless task for a visitor, courteously treated as I was, thus to criticise any institution which the officers permit him to inspect. But I write in the hope of helping, in however humble a way, to bring about a reform in the injurious practice of the State contracting with private individuals for the maintenance of its insane poor. The proprietors receive 11 dollars (45s. 10d.) per head per month for maintenance and clothing. This system involves the probability of patients being sacrificed to the interests of the proprietors. It has the disastrous tendency to keep the dietary as low as possible, to lead to a deficiency in the supply of clothing, and to a minimum of attendants, thus inducing a want of proper attention to the patients and an excessive resort to mechanical restraint, instead of that individual personal care which is so needful for their happiness and the promotion of their recovery. I consider that the number of attendants in such an asylum should not be less than 1 to 8, instead of 1 to 15; and that a higher class should be obtained by giving higher wages. At present they are as follows: £1 16s. to £2 a month in winter for male attendants; £2 8s. to £2 16s. in summer. Women attendants have 5 to 6 dollars a month, or £12 to £15 a year. With a higher class, it might no longer be an irony to speak, as the chaplain does in one of the Reports, of "the good and virtuous keepers who are selected with great discernment."

I venture also to express the earnest hope that the Province of Quebec will itself undertake the responsibility of providing the necessary accommodation for its insane poor and their skilful treatment, that a resident medical superintendent, with full authority, will be appointed, and that there will be a Board of Management, as well as really efficient inspectors.

Since this article was written, the following has appeared in the "Canada Medical and Surgical Journal":—

"At a largely attended meeting of the Medico-Chirurgical Society of Montreal, held on November 7th, 1884, the following resolutions were unanimously passed:—

"1. That this Society has every reason to believe that the statements contained in the Report of Dr. D. Hack Tuke, of London, England, upon our Provincial Lunatic Asylums, are, in every material respect, true and well-founded.

"2. That these statements show a most lamentable state of things as regards the general, and especially the medical, management of these Institutions.

"3. That it appears to this Society to be the imperative duty of the Provincial Government to institute a thorough investigation by competent persons into the entire system of management of the insane poor in this Province.

"4. That the 'farming' or 'contract' system, either by private individuals or by private corporations, has been everywhere practically abandoned, as being prejudicial to the best interests of the insane, and producing the minimum of cures.

"5. That in the opinion of this Society all establishments for the treatment of the insane should be owned, directed, controlled and supervised by the Government itself, without the intervention of any intermediate party.

"6. That the degree of restraint known to be employed in our provincial asylums is, according to the views of the best modern authorities, excessive. That the ablest European, American, and also Canadian alienists have almost entirely given up any method of mechanical restraint. That these facts call urgently, in the name of humanity, for reform in this direction in our provincial asylums.

"That this Society concurs fully in the opinion already expressed by Dr. Tuke in his Report,to the effect that 'the authority of the Visiting Physician (Dr. Henry Howard), appointed and paid by the Government, has been hitherto almost, if not entirely, *nil*. His hands have been so tied that he could not be held responsible for the way in which the asylum has been managed.'"*

---

* A Bill has been since introduced into the Legislature which, when it comes into force, will be of some service. Nothing, however, but a radical change in the system of authority and the mode of maintenance will suffice. I am glad to hear that a new asylum, established on sound principles, is to be erected near Montreal.

The Insane in Ontario.

There were, on the 30th September, 1883, 2,825 patients in the Provincial Asylums of Ontario. This is an increase of 83, or 3·02 per cent. over the previous year. There were two less in the Insane Wards of the Kingston Penitentiary; and the insane in jails, awaiting removal, were fewer, viz., 34 instead of 47. There were 23 patients at home on trial. In all there were 3,070 insane and idiotic persons officially recognised, being 137 more than in the previous year.

They were thus distributed:—

	M.	F.	Total.
Toronto Asylum	358	345	703
London Asylum	440	455	895
Kingston Asylum	230	219	449
Hamilton Asylum	246	301	547
Total insane in Asylums	1,274	1,320	2,594
Asylum for Idiots at Orillia*	122	109	231
Total	1,396	1,429	2,825
Insane Convicts in Kingston Penitentiary	29	2	31
Insane Idiots in Common Jails	21	13	34
Total	1,446	1,444	2,890

If to these numbers are added the patients whose names are on the files for admission into the above asylums, viz., 157, and the number of patients out on probation, viz., 23, we obtain the total number known to the authorities at the above date, viz., 3,070.

Dr. O'Reilly, of Toronto, the Inspector of Asylums, who provided me with this information, states that there were 2,837 beds in the asylums of the Province; so that

---

* On Lake Simcoe. Dr. Beaton is the superintendent.

as 2,825 patients were resident at the time of this return, and a certain number were out on trial, it is clear that at that period the capacity of the institutions was pretty fully reached. More recently, additional buildings have been erected, but as the number of insane has increased, the relative proportion of supply and demand is probably about the same. It may be stated that the admissions during the year ending September, 1883, were, as regards the asylums mentioned, 543; the number discharged recovered was 174; improved, 52; and the number of deaths, 183. The ratio of recoveries to admissions, viz., 33·52 per cent., is stated by Dr. O'Reilly to be higher than in any year since 1877. The idiots are, of course, excluded. The mortality, calculated upon the average number resident, was 6·31 per cent., which is lower than for some years previous.

The total annual cost per patient in 1883, in the four asylums for the insane, varied from 127 dollars 16 cents (Hamilton) to 145 dollars 12 cents (London); the weekly cost being, respectively, 2 dollars 44 cents and 2 dollars 79 cents. Dr. O'Reilly contrasts the low rate of expenditure in the Canadian asylums with that of the United States, where the lowest average is 227 dollars 75 cents per annum per patient; while in Ontario the average is 134 dollars 68 cents. And he quotes the saying of one of the superintendents of the asylums in Ontario that this scale is "nearly poor-house rates." The same contrast has struck me as very remarkable, and the explanation is not altogether creditable to the Canadian Government. The salaries of attendants and servants are lower in Canada, but the explanation of the difference is to be found, according to the Inspector, in the relative character of the lodging, clothing, and, he proceeds to say, more especially the food. The quality is said to be good, but it is "plain and unattractive," so as to become extremely distasteful to many patients. It is difficult for a stranger to form an opinion on this subject, because he is unacquainted with their diet at home; but the asylum dietary is as good as in our county asylums.

The revenue from paying patients, of whom there were 538 in the asylums, amounts to a very considerable sum, viz., 59,922 dollars (£12,485) during the last year. This certainly points to the probable success of a private asylum which has recently been established at Guelph, of which Dr. Lett is the superintendent.

I find, from a return made of the number of patients employed in the asylums of Ontario, that 52·57 per cent. of the patients were engaged in some occupation; being 1,479 out of an average population of 2,813. The largest percentage was at the London Asylum, viz., 69·89.

The authorities in Ontario are not blind to the difficulties connected with the accumulation of incurable patients, for whom the question of separate accommodation arises. I am glad to observe that the latter difficulty is being met by the erection of small buildings; these being sometimes devoted to the curable class of cases, while the larger buildings are retained for the incurable. In some instances, however, small buildings or cottages are, and may properly be, devoted to the chronic insane; while the recent cases are treated in the old and more expensive building. Dr. O'Reilly believes that the general feeling is entirely in favour of detached wards or annexes, and I certainly found this to be the case at the asylums which I visited.

It will be advantageous to state here a few of the leading provisions of the existing statute relative to lunatics, enacted in 1871 by the Legislative Assembly of the Province of Ontario, entitled "An Act Respecting Lunatic Asylums and the Custody of Insane Persons" (cap. 220).

The Public Asylums are established and acquired under a grant from the Legislature of the Province, and are vested in the Crown.

The Lieutenant-Governor has the appointment of the Medical Superintendent.

Among the duties of the Medical Superintendent are those of reporting the condition of the asylum to the Inspector of Prisons and Public Charities at each visit, and also of reporting annually to the Inspector upon the affairs of the institution. The financial affairs of asylums are conducted by the Bursar, who is appointed by the Lieutenant-Governor.

The salaries of these officers are fixed by the Lieutenant-Governor, and do not exceed £400 for the Superintendent, and £240 for the Bursar.

In regard to admissions, no patient can be admitted (except upon an order by the Lieutenant-Governor) without the certificates of two medical practitioners, each attested by two witnesses, and bearing date within three months of admission. Each certificate must state that the examination was made separately* from any other practitioner, and after due inquiry into all necessary facts; the medical practitioner specifying the facts upon which he has formed his opinion, and distinguishing those observed by himself from those communicated to him by others. Dangerous lunatics may be committed to jail by a Justice's warrant on his receiving the necessary information, and after evidence given with reference to the prisoner's state of mind. He remains in jail until removed to an asylum by the Lieutenant-Governor, where he remains until discharged by the same authority.

The Inspector of Public Charities is *ex-officio* the Committee of every lunatic having no other, who is detained in any public asylum of the Province. The Court of Chancery may at any time appoint a Committee of any such lunatic, if it considers it expedient, in place of the Inspector. The Chancellor, who may call experts to his assistance, decides the question of mental unsoundness and incapacity to manage his affairs, without a jury. I understood that the number under the legal guardian-

---

* Curiously enough the previous Act required the examination to be made by the physicians together. "Three months" is a long period.

ship of the Chancellor is somewhat under 400. They are placed in confinement under his warrant. For this class, the legal checks are much more stringent. They are subjected to more official recognition, and they cannot be discharged without the sanction of the Inspector. The ordinary patient, whatever may be his social position, is admitted into an asylum on two medical certificates, and he may be discharged by the Superintendent without reference to the Inspector.

In regard to *Private Lunatic Asylums,* Justices of the Peace assembled in general sessions may grant a license to any person to keep a house for the reception of lunatics within the county. The regulations for private asylums are moulded upon those of the English Lunacy Laws, and need not, therefore, be given.

Returns are made monthly by the asylums to the Inspector in regard to admissions, discharges, and deaths. It is obvious that if these returns are made with a view to prevent improper admissions, or to allow of an inquiry in alleged deaths from violence, far too long a period elapses before the Inspector has cognizance of an admission or death at an asylum. It is argued that in the case of the private institution at Guelph, a Committee has general oversight over it, and that this constitutes a sufficient guarantee against abuses. But however good it may be, and doubtless is, it does not supersede the necessity of an independent Government official receiving immediate information in regard to the admissions and deaths of patients in every asylum, for the inspection of which he is responsible to the public. And, before dismissing the subject of inspection, I would say it is a great defect in the law which enacted it, that it is not made imperative to have one of the Commissioners a medical man.

Passing to the asylums themselves, I will first refer to the asylum at *Toronto,* which I visited on the 20th of August last. The contrast, as I have elsewhere intimated, between the asylums of the Province of Ontario and those of Quebec is really astonishing. The system is essentially different. The Legislature of Ontario recognises the duty not merely of discovering institutions to which it can send its insane poor at so much a head, but of providing the institutions themselves, and making the State responsible for their proper management. I do not maintain that all has been done that can be done, or in all instances on a sufficiently liberal scale, nor yet that the asylums are perfect in their organization and management, still less that the system of inspection is the best that can be devised; and I object to any alleged dangerous lunatic who has not committed a crime being in the first instance sent to jail, and thereby branded as a criminal: but I have no hesitation in saying that there is a sincere endeavour to make adequate provision for the insane of the Province; that the inmates of the asylums are carefully treated, and that there exists among the superintendents a real interest in their work, and a desire to do their duty to their patients.

At the Toronto Asylum, superintended by an active administrator (Dr. Daniel Clark), there are 710 patients, the sexes being almost equal. The spacious corridors (15 feet in width) and rooms are carpeted, and altogether well furnished, and in those used by the destructive patients there is not the dismal bareness too often witnessed. There is strong evidence of the great care and attention paid in this asylum to cleanliness, and to the dress and general comfort of the inmates. There was hardly anything deserving the name of mechanical restraint. On the female side there has been practically none for two years, and as regards the men patients there has been none whatever, Dr. Clark informed me, for seven years. No patient was in seclusion at the time of my visit. Indeed, Dr. Clark strongly objects to its use. There is one feature in the construction of the asylum which attracts the notice of the visitor at first sight, not very pleasantly, it must be admitted, and that is the succession of semi-circular spaces or verandahs at the end of the corridors, protected and enclosed as they are by strong iron palisades. A glazed wooden frame partitions off these spaces from the corridor. In the areas of these projecting spaces the patients stand or sit on chairs, gazing on the outer world through the vertical bars. On those who look up to them from below, the impression of a cage in a zoological garden may be, and indeed has been, produced. At the same time, it is surely much better for the patients to be able to step outside the corridor into such an enclosure and breathe the fresh air, than not. The view over the Lake (Ontario), etc., is extensive, and affords variety, while the objection which may be made in regard to the effect produced upon other minds is rather sentimental than practical. In a new building no doubt this particular construction would be avoided, or an ornamental guard would be constructed in place of simple bars.

The pay of the attendants, with whose appearance I was pleased, both as regards personal expression and dress, is liberal—£3 15s. to £5 6s. a month for males, and £2 to £2 8s. for females. In the wings there is one attendant to 12 patients; not so many in the central large wards. There are also six night watches, three on each side of the house. There are six galleries for private patients. They pay from three to six dollars a week. There are also six free wards. Four hundred patients pay nothing. The weekly cost per patient is a little more than $2\frac{1}{2}$ dollars a week, or 134 dollars (£27 6s.) a year, exclusive of the capital account or repairs.

The patients are employed to a considerable extent, namely, about 60 per cent. of the free class, from whom alone work can be obtained. All the vegetables required for the asylum are raised on the grounds. There are 140 acres. Dr. Clark, however, states, in the report he favoured me with, that the last potato crop had proved a failure, but that the other crops were about the average.

As there are about 29 acres under crop, the potato failure was a serious one for the asylum. As there are no crops of hay and oats, the cultivation of roots is mainly attended to, and Dr. Clark calls attention to the need of more arable land. The value of the produce of the present small farm was 13,763 dollars in 1883. Buildings, including a prison, have grown up in the vicinity; a regrettable circumstance, especially if, as I understand, land belonging to the institution has been sold for building purposes. There are, distinct from the main buildings, three cottages, in which 120 female and 50 male patients are accommodated. One is cheaply built, and is well adapted for the purpose. There are good day-rooms and dormitories. The floors are partly carpeted, and there are a few pictures on the walls.

The separation of cases, which these annexes allow, affords advantages which here, as elsewhere, are fully appreciated.

This, as well as the other Ontario Asylums, is inspected by one of two Inspectors of Public Charities and Prisons in the Province. He visits four or five times in the course of the year, and oftener if he sees fit. The Grand Jury have the power of visiting the asylum if they wish, and when they do so they make a presentment to the Court. Their visits, however, are, I believe, of a somewhat formal character.

This asylum was opened in 1843, and was at that time the only institution for the insane in the Province. Indeed, this was the case when the well-known and universally esteemed Dr. Workman became superintendent in 1853. At that time there were only 300 patients. What the condition of the asylum was, two years after it was opened (and I have reason to believe up to the time Dr. Workman became superintendent) I have the means of stating, on the authority of my brother, Mr. J. H. Tuke, who, on visiting it in 1845, made the following entry in his diary :—

"*Toronto, Sept. 30, 1845.*—Visited the lunatic asylum. It is one of the most painful and distressing places I ever visited. The house has a terribly dark aspect within and without, and was intended for a prison. There were, perhaps, 70 patients, upon whose faces misery, starvation, and suffering were indelibly impressed. The doctor pursues the exploded system of constantly cupping, bleeding, blistering, and purging his patients; giving them also the smallest quantity of food, and that of the poorest quality. No meat is allowed.

"The temples and necks of the patients were nearly all scarred with the marks of former cuppings, or were bandaged from the effects of more recent ones. Many patients were suffering from sore legs, or from blisters on their backs and legs. Everyone looked emaciated and wretched. Strongly-built men were shrunk to skeletons, and poor idiots were lying on their beds motionless, and as if half-dead. Every patient has his or her head shaved. One miserable court-yard was the only airing-court for the 60 or 70 patients—men or women. The doctor, in response to my questions, and evident disgust, persisted that his was the only method of treating lunatics, and boasted that he employs *no restraint*, and that his cures are larger than those in any English or Continental asylum! I left the place sickened with disgust, and could hardly sleep at night, as the images of the suffering patients kept floating before my mind's eye in all the horrors of the revolting scenes I had witnessed."

Dr. Workman reformed the asylum; and, could an unvarnished tale be told of the condition in which he found, and of that in which he left it, no better tribute could be paid to his character and work during the period of his superintendence.

Dr. Workman now resides at Toronto, and has attained to nearly 80 years of age. His mind is still extraordinarily active; and his pen is frequently in his hand, engaged in both original writing and in making translations from foreign medical journals. As longevity is in the family, it is no mere form to express the hope that this Nestor of Canadian specialists may pursue his literary work for many years to come. In making Dr. Workman an Honorary Member of the Medico-Psychological Association, the latter honoured itself as well as him. In conversing with me on the provision required for the insane in Ontario, he gave it as his decided opinion that there had been an increase in their number, beyond what either the increase of population or the accumulation of chronic cases could explain. Although the proportion of ascertained lunatics is about 1 to 700, Dr. Workman estimates that there is in reality 1 to 500. Formerly there was no general paralysis, now it is common enough; not so common, however, as in England, for at the Toronto Asylum there were not, at the time of my visit, more than a dozen cases; and there are only three or four deaths from this disease in the course of a year. Dr. Clark considers it more frequent among the higher classes than among the lower.

I visited with much interest the *London Asylum*, which Dr. Bucke superintends with great energy and enthusiasm. Not only is the town itself called after London, but the river upon which it stands is the Thames; and it boasts of its Westminster Bridge and its Piccadilly. The resemblance does not end here: for if it be allowed that there is a good asylum in or near our Metropolis, it will not be denied by anyone who inspects Dr. Bucke's institution, that its analogue resembles it in this particular also. It was opened in 1870, and the present superintendent entered on his duties in 1877. The whole establishment, the main building, the separate one for the refractory patients, the cottages and the farm, convey the impression of active life, and of the sustained interest of an able head. Dr. Bucke has resolutely set himself to employ the patients in some way or other, especially on the farm

—with great advantage, it need scarcely be said, to their mental and bodily health, and with the result of emptying the wards of those helpless, hopeless, cases whose drear existence in the dead-alive asylums of any country suggests *cui bono* to the pessimist, and makes even the optimist sad at heart. If Dr. Bucke is asked how he employs a man in a state of acute mania, he replies, " Oh, I make him break stones."

Without taking the reply too literally, it may serve to show the exceeding but just importance attached to labour or being out of doors, which has been so long and frequently maintained in the Mother Country. I gathered from my enquiries that very few cases of mania with exhaustion are admitted to this asylum, a very important fact in this connection, and one which might have been expected as a point of contrast between the admissions into the asylums in old and new London respectively. Mania in some form is about four times as frequent as melancholia. Only one patient was instrumentally fed last year.

The number of patients in this asylum is 888; 438 males, and 450 females. It has a capacity for 906. The estate consists of 300 acres, 200 being occupied by the farm, and 40 by the gardens, while the buildings cover a large portion of the remainder.

The main building cost a little more than £100 per bed. (Land is here about £30 an acre.) It accommodates about 500 patients of both sexes, of the quieter class; and an assistant medical officer, Dr. Burgess, resides here. It consists of the usual arrangements—corridor (12ft. in width), recess, day, and bedrooms, some of which latter are dormitories containing 16 beds. The number of single rooms in the whole establishment is 250. As I went through the men's side, as many as 250 patients were at dinner in an associated dining-room. All had meat, and I found this was usual.

There is a distinct three-storied building for patients of

a more or less excited character, male and female. The first assistant physician, Dr. Beemer, resides here. There is nothing special in the arrangement of the wards. There are 184 single bedrooms, affording 720 cubic feet of breathing space per patient. The windows were unnecessarily guarded by iron bars and net-work. No doubt these are survivals of the past, and if rooms for the refractory were now built at this asylum, no trebly guarded window would be introduced, for it is out of character with the air of freedom which everywhere prevails in the institution. More light would be also admitted into the building. There is a good airing court, shaded by trees, and provided with a shed and seats. In this asylum, as in most others on the other side of the Atlantic, the number of epileptics is small—only about 25. There was no patient in restraint, and none in seclusion. Dr. Bucke observed that it was rare to have black eyes among the patients since he determined not to resort to mechanical restraint unless absolutely necessary. No patients were crouching on the floors in strong dresses. I must add that " chemical restraint " is not resorted to in the asylum. Sedatives are rarely given, even in small doses. In addition to the morning round, I went through the wards after the patients were in bed, and there was very little noise indeed. Before quitting this building for the excited patients, I should state that, of 92 men residing in it, from 75 to 80 are on an average employed.

Dr. Bucke observes in his last report: "The disuse of all forms of restraint, and the employment of so large a proportion of the patients in the asylum, have been accompanied by (or have caused) an unmistakable elevation of the tone of life throughout the whole institution; and as one evidence of the fact, I may mention that the Sunday attendance at chapel has nearly doubled during the year just closed; a year ago the average attendance at Divine service on Sunday morning being about 260, and now over 400. . . . Along with the disuse of restraint and seclusion, we have almost entirely ceased using strong dresses, of which, up to within the last few months, we were in the habit of using a large number, and although we now use no restraint or seclusion, and hardly any strong dresses, we have less tearing of clothes and bedding, and breaking of furniture, etc., and far less striking and fighting on the part of the patients than when restraint and strong dresses were freely used. It should also be mentioned that we use absolutely no sedatives of any kind; and it is seldom indeed that any patient is held or restrained, even for a few minutes at a time, by the hands of attendants. The last fact was a very surprising one to me, for I had always believed that when mechanical restraint was discontinued in any asylum, manual restraint had to be substituted for it; and the chief argument which I have in former times used, and heard used, against the discontinuance of mechanical restraint, has always been that it was much preferable to restraint by the hand of an attendant, always wrongly taking it for granted that where the former was not used, the latter must be."

In addition to the main building and the north or refractory branch, there are two excellent but cheaply constructed brick cottages, containing 60 patients each. The cost amounted to 32,000 dollars, or about £55 per bed. The patients in these cottages are either convalescent or able to appreciate the comparative independence of a separate house, not presenting any appearance of an asylum for the insane. The rooms were tastefully furnished and very clean.

There is still another cottage for 60 male patients—those who are particularly engaged in working on the farm. The cost was 18,000 dollars, being at the rate of a little more than £60 per bed.

As compared with most County Asylums in England, the furnishing of the main and north buildings struck me as somewhat scant. I was told that the patients of the class that go to the London Asylum are not accustomed to more, at home, in the way of carpets, &c., than they find when they come to the asylum. It is true, also, that they are so much out of doors that they may not care much for somewhat bare corridors and rooms. The cost per head for maintenance amounts to 105 dollars 12 cents, or about £22 a year; this includes, in addition to food, salaries and furniture, but not any considerable repairs or the additional buildings—certainly a low figure—and it should be mentioned that about 80 per cent. of the patients are clothed by the institution. I have already said that the total cost per annum of patients at the London Asylum amounts to 145 dollars 12 cents, or 2 dollars 79 cents per week. The above charge for maintenance is no doubt kept down by the large yield from the farm and garden, although the total cost is greater than in any asylum in Ontario. I wished to ascertain the exact extent of this, but the accounts at the Superintendent's command did not show it, nor was the Inspector, Dr. O'Reilly, able to put me in the way of obtaining this information, valuable and interesting as he felt it to be. A clear estimate of the net profit would greatly redound, I doubt not, to the credit of the institution and of the strenuous endeavour made to have a profitable farm connected with an asylum for the insane. Dr. Bucke drove me over the farm. Its produce and that of the gardens were roughly estimated by him at about £3,000 a year. There were 200 pigs on the day I was there. Over 100 are killed every year. Some 6,000 bushels of potatoes are raised annually, and as many quarts of berries from the gardens. Last year the crop of hay amounted to 140 tons. The asylum has 40 cows.

As none of the patients pay a cent (for it is a genuine pauper asylum), it is doubtless easier to induce them to work than in mixed institutions, and also to find men accustomed to farm, and to be handy at various trades. To compare the amount of work done at such an institution with one for private or mixed patients would be very unfair. It will not, however, be denied that there are some pauper institutions in the world in which the patients do little or no work from year's end to year's end, and spend a much larger proportion of the day in the wards of the asylum than out of them. Nor is it altogether impossible that there are institutions of a mixed class in which the patients might do a little more work both indoors and out, especially the latter, than they do already. In this I include the constant attempt to induce the patients to take exercise in the open air with as definite an object as possible. This can only be effectually done by a superintendent who has his heart in the work, and who will insist upon having a sufficient staff of attendants, even on the score of economy, should those who hold the purse-strings be deaf to an appeal to higher motives. But what if there is no breathing space outside the walls of the asylum? Then, woe betide the superintendent and the unhappy patients under his care. Their fate is sealed.

On examining the record of work, and taking a single day, I found that out of the 438 men no less than 392 were employed; while out of 451 women, 404 were occupied in some sort of work. Of 40 that do not work, 25 are physically incapable, and 15 cannot be induced to work without more pressure than it is thought right to use. I am well aware that figures like the above may mean much or little, but I am satisfied from personal observation that in this instance they mean much.

It is especially interesting to observe how a better system of treatment has become possible by the increased employment of the patients. With 880 patients the average number at work was, at the date of Dr. Bucke's last report, 625. He observes: "I have always found that, no odds how violent a patient is, if you can once get him or her to work, the case will give you very little further trouble in that way. . . . The male patients have been engaged in all the various kinds of farm and garden work; they work with the carpenter, mason, painter, tailor, engineer, baker and butcher; they work in the horse and cow stables, and do most of the milking; they assist in dining-room, kitchen and laundry; they sew, knit, make and mend shoes, boots, and slippers; seat chairs with cane and reed; make mats; they do tinsmithing, blacksmithing, locksmithing, upholstering, clerking; all kinds of work in the halls, as bed-making, sweeping, scrubbing; sawing and splitting wood, shovelling coal, grading land, making roads, feeding and tending two hundred pigs, working in the store, picking hair for mattresses, and doing all sorts of odd jobs. The female patients are largely engaged in sewing and knitting; and, besides, they work in the kitchen, laundry, and dining-rooms; do all sorts of work in the halls, as bed-making, sweeping and scrubbing; milk, pick hair for mattresses, and gather fruit and vegetables in the gardens."

The proportion of attendants to patients is certainly not high in the London Asylum; in fact, the Province ought not to complain if the Superintendent should increase the number. For the violent patients, the proportion was one in nine; for the others considerably less. It ought, however, to be remembered that the number of ward-attendants does not adequately represent the services rendered to the patients, inasmuch as those workmen who labour on the grounds or at any handicraft, exercise surveillance over some of the patients at the same time. Several years ago, Dr. Eames (Cork Asylum), the President of the Medico-Psychological Association, urged upon his Committee the need of more attendants, and stated that while the proportion of attendants, with the above-mentioned helps, was one to eleven in his asylum, it averaged about one to eight in the asylums of Ireland generally. The maximum pay of male attendants at the London Asylum is about £50 a-year; that of the females is about £30. On the male side are several female attendants—not the wives of attendants, as at Brookwood and some other asylums in England, but respectable widows. Dr. Bucke attaches great importance to this feature of his management, as ensuring cleanliness, tidiness, and consideration. He states that he has no difficulty in finding suitable persons. He is fortunate, for he requires pleasant manners, industrious habits, good feeling, and, above all, good sense, in addition to widowhood. They must be widows indeed. To do him full justice, however, I must cite a few passages from his last annual report:—

"The first was engaged in January, 1883, and became the supervisor of the upper storey, and does all the work that a man in that position would do, and besides that, she has a general supervision over the tidiness and cleanliness of the whole wing; the other two women act as her assistants on the other two flats. They look after the men's clothing, see to the tidiness of the beds, cleanness of the floors, &c., &c.; and especially, they oversee the indoor work of a large number of male patients, who pick hair, sew, knit, make mats, &c. But the active duties of these women, though important, are scarcely so valuable as is their mere presence in the halls, which has a strong tendency to check improper and unseemly talk and conduct, so that these halls are different places now from what they used to be before these women took service in them. . . . Down to the present time, none of them have been by speech or action either injured or insulted by any patient. Almost universally the patients like to have them amongst them, and I find that often the women can get the patients to work when the male attendants can get them to do nothing."

In an institution where the gospel of fresh air and employment is fully believed in and carried out, one feels especially interested in the dietary. The meals are taken at 6.30 a.m., 12, and 6 p.m., the patients going to bed after supper up to 9 o'clock p.m. I append the dietary in detail, but must premise that work, whether out or in-door, is not encouraged by the stimulus of beer, for Dr. Bucke is an out-and-out teetotaler. He has not ordered alcohol in any form, even as a medicine, for three years. When he became Superintendent, a considerable sum was expended on beer; more food is now given, but not more milk, which is, I think, to be regretted. The attendants never had any beer, so no money equivalent has been necessary.

The dietary in the main asylum on a particular week which I chose, viz., that beginning June 8th, 1884, was as follows:—

BREAKFAST.

Sunday.—Bread and butter, tea and coffee.
Monday and Wednesday.—Porridge and milk.
Tuesday.—Boiled rice and syrup.
Thursday.—Oatmeal porridge and syrup.*
Friday and Saturday.—Porridge and milk.

DINNER.

Sunday.—Stew, potatoes.
Monday.—Corn-beef, potatoes, and beans.
Tuesday.—Roast beef, potatoes, bread pudding.
Wednesday.—Boiled beef, potatoes, and peas.
Thursday.—Haricot, potatoes, and bread pudding.
Friday.—Fish, boiled beef, pickles, and potatoes.
Saturday.—Roast beef, potatoes, bread pudding.

TEA.

Sunday.—Bread and butter.
Monday and Saturday.—Bread and butter.
Tuesday.—Stewed rhubarb.
Wednesday.—Bread and butter.
Thursday.—Currant rolls.
Friday.—Apple sauce.

With the foregoing may be compared the dietary of an English pauper asylum, that at Hanwell:—

* Molasses.

BREAKFAST.

For males.—Cocoa, bread and butter.
For females.—Tea, bread and butter.

DINNER.

Sunday.—Roast pork, beef, or mutton.
Monday.—Soup, thickened with oatmeal, rice and peas, and containing 2 oz. of meat for each patient; also 6 oz. currant pudding or 10 oz. baked rice pudding.
Tuesday.—Meat pies.
Wednesday.—St. Louis corn beef.
Thursday.—Boiled bacon or pickled pork.
Friday.—Fish, fried or boiled, with melted butter.
Saturday.—Irish stew.

SUPPER.

Tea, bread and butter.

For patients who are employed, luncheon, consisting of bread and cheese and beer (half-pint), is provided in addition; and for Monday's dinner, boiled bacon or pickled pork is given instead of soup.

The following are the principal salaries and wages (mostly with board) allowed at the London Asylum:—

MALES.—Medical superintendent, £420; first assistant physician, £210; second ditto, £210; third ditto, £154; bursar, £290; steward and storekeeper, £166; engineer, £154; two carpenters, £220; tailor, £94; gardener, £83; assistant ditto, £50; butcher, £50; baker, £83; farmer, £125; two ploughmen, £115; cowman, £45; three night-watchmen, £157; three chief attendants, £195; twenty-nine ordinary male attendants, £1,389.

FEMALES.—Matron, £105; assistant ditto (refractory ward), £52 10s.; chief attendant, £52 10s.; thirty ordinary female attendants, £990; three night attendants £82; five cooks and assistant cook, £137; five laundresses, £115; nine housemaids, £195; one dairymaid, £25; two seamstresses, £50.

With the foregoing may be compared the following salaries, &c., at the Hanwell Asylum (750 men, 1,143 women) :—

OFFICERS.—(*a*) Resident medical superintendent (female department), £700 per annum ; (*a*) resident medical superintendent (male department), £700 ; (*d*) chaplain, £350 ; (*f*) clerk to the Committee of Visitors, £275 : (*c*) assistant medical officer, £200 ; (*c*) ditto, £200 ; (*c*) ditto, £165 ; (*c*) ditto, £150 ; (*e*) apothecary, £120 ; (*b*) engineer, £450 ; (*a*) storekeeper, £500 ; (*e*) clerk of the asylum, £325 ; (*e*) first assistant-clerk, £130 ; (*c*) second assistant-clerk, £110 ; (*e*) storekeeper's clerk, £110 ; ditto, £60 ; outdoor inspector, £74 ; (*e*) matron, £345 ; assistant matron and organist, £66 ; junior assistant matron, £40 ; superintendent of laundry, £55 ; superintendent of workroom, £50 ; principal female attendant, £36 ; ditto, £34 ; ditto, £30 ; workroom assistant, £33.

(*a*) Furnished house, rates and taxes free, coals, gas, milk, and vegetables ; (*b*) part ditto, ditto, ditto, washing, milk, and vegetables ; (*c*) furnished apartments, attendance, coals, gas, washing, milk, and vegetables ; (*d*) unfurnished house ; (*e*) dinner daily ; the matron boards two servants ; (*f*) neither boarded nor lodged. The others have board, lodging, and washing.

(*a*) MALE ATTENDANTS.—(*b*) Three supervisors, at £80 ; (*c*) eighteen charge attendants, £25 to £40 ; forty-four ordinary ditto, £25 to £35 ; hall attendant, £40 ; (*d*) six night ditto, £62 to £72.

(*a*) One suit of uniform every eight months, and a suit of serge every two years under certain conditions ; (*b*) do not reside in asylum ; (*c*) have board, lodging, and washing, except in the case of some of the attendants, who are allowed £1 per month in lieu of their lodging and washing. Three out of the eighteen receive £47. (*d*) These attendants are neither boarded nor lodged.

(*a*) FEMALE ATTENDANTS.—Four supervisors, £30 to £39 ; twenty-five charge-attendants, £15 to £29 ; eleven night ditto, £19 to £32 ; seventy-one ordinary attendants, £15 to £25.

(*a*) All have board, lodging, and washing ; and suits of uniform every eight months.

(*a*) KITCHEN AND LAUNDRY.—One head cook, £46; one assistant ditto, £26, with three suits of uniform every eight months; one ditto, £20, with ditto; two kitchen-maids, £14 to £20, with ditto; one head laundress, £20 to £25, with ditto; one assistant ditto, £18 to £25, with ditto; one officers' ditto, £18 to £25, with ditto; eleven laundry maids, £15 to £25, with ditto; seven domestic servants, £14 to £20.

(*a*) All have board, lodging, and washing.

WORKMEN.—One upholsterer, £1 12s. per week; one ditto, £1 4s.; two ditto, £1 and 18s.; two tailors, £1 8s. and £1 4s.; one tailor, £1; two shoemakers, £1 10s. and £1 3s.; one tinman, £1 9s.; one basket-maker, £1 7s.; (*a*) one butcher, £1 4s.; (*b*) two bakers, £1 6s. and 17s.; (*c*) one gardener, £1 16s.; one ditto, £1 4s.; (*d*) one carter and driver, £1; one carter, £1; one cowman, £1 1s.; one ditto, 18s.; one gardener (front grounds), £1 1s.

(*a*) Breakfast and dinner daily; (*b*) boarded, &c.; (*c*) allowed vegetables; (*d*) lodged and allowed coals, gas, milk, vegetables, and beer. All have an allowance of beer.

I next visited the *Hamilton* Asylum. This institution, opened in 1875, is beautifully situated, overlooking Lake Ontario at the point of Burlington Bay. The situation, however, is not altogether advantageous. It is inconveniently near a precipitous descent, and the approach to the asylum is troublesomely steep. It was originally designed for an inebriate asylum, but the needs of the insane were justly deemed more pressing and practical than those of dipsomaniacs.

Dr. Wallace is the medical superintendent. Unfortunately he has been out of health for some time, for which he has had to travel abroad, but he is now much stronger.

There are 567 patients in the house, of whom 270 are males and 297 females. About 5 per cent. of the patients pay, but only from 6s. to 10s. a week. The construction of the building is on the ordinary linear plan; it is a handsome structure. The superintendent's house is distinct from, but close to, the institution. When I was going round, a number of patients of both sexes were dining together—105 men and 95 women. The dietary was good. The heating and ventilation of the house, the former by steam and the latter by flues to the roof, are well secured.

In the refractory galleries, the least excited patients are, I was glad to observe, placed in the upper storey. Frequently in asylums on the American Continent, the most violent are placed at the very top of the house, a practice very likely to involve neglect and the omission of proper outdoor exercise. The bringing of this class of patients up and down-stairs is in itself a frequent cause of outbursts of excitement and struggling.

With regard to restraint, Dr. Wallace informed me that when he regarded it necessary, he employed leather muffs for the men and the camisole for the women. Were a patient actually suicidal, he would at night, if not in the day, be placed in restraint, a trustworthy patient being placed in the same room. Some months had elapsed since a male patient had been restrained. At the time of my visit a woman was in restraint, who persistently mutilated her face. When the camisole was removed she immediately resumed her injurious work. Judging from the reports of the Inspector, I should conclude that there has been a remarkable diminution of restraint during the last few years.

On the female side there is a sewing-room, where many of the patients work. All the sewing required by the institution is done here. As I am speaking of employment, I may add that for the male patients it is found convenient, in addition to other work, to employ them in winter (it being more difficult then to supply employment) in breaking stones under a shed.

The following is a statement of the employment of patients during the quarter ending June 30th, 1884:—

FEMALE PATIENTS.

Nature of Employment.	Number of Patients Working	Number of Days Worked.
Laundry	15	1170
Kitchen	7	546
Sewing-room	12	936
Dining-room	13	1014
Mending	6	468
General Work	5	390
Knitting	30	2340
Work in Halls	35	2745
Store-room	8	624
Total	131	10233

MALE PATIENTS.

Nature of Employment.	Number of Patients Working.	Number of Days Worked.
Laundry	5	390
Kitchen	5	455
Tailor's Shop	2	147
Dining-rooms	8	728
Carpenters	6	292
Engineers	4	281
Masonry	12	465
Roads	12	322
Coal and Wood	20	849
Bakery	1	78
Dairy	6	546
Butcher	1	91
Piggery	2	182
Painting	2	112
Farm	12	524
Garden	20	745
Grounds	6	211
Stable	1	91
Halls	60	5460
Store-room	2	156
General Work	25	1652
Quarry	50	1400
Total	262	15177

In the refractory galleries on the men's side, the number of the attendants was certainly too small. However, not only was no patient in restraint, but none were in seclusion or in a strong dress. A separate building for a certain number of the refractory class has been erected, and will be shortly occupied. This is another illustration of the tendency there is to adopt the plan of separation of classes of patients, which has been carried out for some years in Great Britain. It is a neat red-brick building, with a limestone basement, and consists of a centre and two wings, having two storeys. It will accommodate 60 men. The cost seems high compared with some of the separate buildings which I have mentioned, viz., £120 a bed, but this is due to the class of cases for which the building is designed being acute instead of chronic. There are rooms on both sides of the corridors. The single rooms are well adapted for their purpose, but the provision for ventilation appears to be scarcely sufficient. The construction of the building readily admits of separating the most from the least noisy patients, and also for placing patients, on admission, under special observation, if desirable. When the building is occupied, an assistant medical officer is to be resident in this building. He has not yet been appointed. Should a false economy prevent his appointment, the separation of this, the most important, class of the insane from the rest of the household, still further removed as they will be from the superintendent's quarters, will be an evil instead of a blessing. That such an evil is not imaginary, I can assert from what I have witnessed in those Continental asylums, in which the paramount idea seems to be to remove violent and dirty patients as far as possible from the centre of the asylum, and that without any medical officer.

There are objections, doubtless, to placing maniacs close to the central offices; but of the two evils I am sure that, for the interests of the patients, to whom every other consideration ought to be sacrificed, this arrangement is better than putting them beyond the reach of sound and sight. I was glad to find that at the Hamilton Asylum an assistant medical officer resides in the main building, near the wards for the refractory male patients. It is to be regretted that this is not the case in every institution for the insane, (unless, better still, one assistant is on duty all night.) He ought to be cognisant of noise if it is unusual, and to be within easy call. It will be said that the appointment of night-watches renders abuses or neglect impossible. This I entirely deny. No asylum is free from the possible, or rather probable, ill-treatment of patients when out of sight of the heads of the institution; but at no time is this so likely to occur as with the violent class during the night and early morning, for then it is that the patients and their attendants are least under observation.

There is another cottage on the ground, which was, I understand, formerly occupied by the bursar. This is now occupied by 19 female patients of a harmless kind. It looked home-like and clean, and the inmates, who were quite of a humble class, seemed very comfortable and contented. This cheerful cottage might be used for the convalescent class. It is comparatively inexpensive.

The attention paid to the dirty patients is highly creditable. The night-watches carry out the system of getting this class up to the fullest possible extent. I looked at the reports handed in to the superintendent in the morning, and found the number of beds reported soiled remarkably small. On the day I was at the asylum there were only two on the female and one on the male side. There are four night-attendants. I also examined many of the beds when passing through the dormitories, as also did Dr. Ashe, of the Dundrum Asylum, who happened to join us in our round, and we were struck with the cleanliness of the bed-linen in the division where it was most likely to be foul. I may state that only five men in this asylum were the subjects of paresis, and two women. Hence, as compared with an asylum of the same size in England, the number of patients likely to be dirty would be much smaller.

No alcohol is used at this asylum except medicinally, and that rarely. Formerly beer was an article of diet. Since its disuse, milk has been given as a substitute if the patients desire it. No money equivalent was given to the attendants. Their salaries reach £50 per annum for men and £25 for women.

The last asylum I visited in Ontario was that at *Kingston*. In the absence of the superintendent, Dr. Metcalf, his brother-in-law, Dr. Clarke, the assistant medical officer, obligingly showed me over the institution. It is situated on the north bank of Lake Ontario. There are 255 male, and 250 female patients. These 505 patients are paupers, with the exception of a very few who pay the cost of maintenance, viz., two dollars, or nearly 8s. 6d. a week. The asylum, which is of stone, was opened in 1859. It is built in the usual corridor style, and has four storeys in addition to the basement, which is not used for the patients. There are 180 single rooms, 90 of which are for the worst class. The associated dormitories have not more than 11 beds in any one of them. The breathing space per patient amounts to 1,034 cubic feet in the former, and 700 in the latter. In this asylum, the suicidal patients are distributed in dormitories, with other patients on whom reliance can, to a considerable extent, be placed. In addition, the attendants' door opens into the dormitory, and the night-watch looks in every hour. There has been no suicide since 1877.

The estate covers 140 acres, 85 being devoted to the farm and garden, on which patients are employed. Eight patients look after the cattle; 25 work on the roads; five assist the engineers; two are carpenters, two painters, three tailors, two shoemakers, two bakers; two assist in the kitchen; and 160 are employed in the wards. Of the women, upwards of 150 are employed.

I was glad to see here, as at the other asylums in Ontario, cottages for certain classes of cases. One cottage was occupied by 37 women of the quiet and incurable class. An annexe, only opened this year, for 70 patients of both sexes, and built of limestone, cost 30,000 dollars, including warming apparatus and furnishing, or about £100 per bed. There are no single rooms in the house. The centre consists of four, and the wings of three storeys. At the present time the annexe is full.

The general appearance of the patients at this asylum was very satisfactory. Evidently they are under kind and skilful management. The asylum is inspected four times a year by Dr. O'Reilly, and nominally by the Grand Jury at the Assizes.

It is a matter of some interest to be able to compare the salaries given to the staff in an asylum in Canada with those allowed in England. For this purpose I append the salaries of the officers at the Kingston Asylum:—

Medical superintendent, £333, with house, rations for himself and family, &c., &c.; assistant medical officer, £210, with like extras; bursar, £240, dinner on the premises; steward, £100, with house, rations, &c.; storekeeper, £100, dinner on the premises; engineer, £155, with house and garden; assistant-engineer, £83, with meals and lodging, and stoker, £60, with meals and lodging; farmer, £72, with house, garden, and meals; gardener, £83, with house and garden; ditto vegetable garden, same; butcher, £50, with house, garden, and meals; baker, £83, with like extras; tailor, £83, with meals; carpenter, £90, with ditto; the night-watch (male side), £72, with meals; female night-watch, £50, with meals; head male attendant, £83; ten attendants, £72 each, and eight attendants, £50; matron, £83, with rooms, rations, &c.; assistant-matron, £41 10s, with like extras; thirteen female attendants, £25; two night-watches, ditto; laundress, £30, with meals; assistant-laundress, £25; cook, £30; under-cook, £25; dairymaid, £20; two domestics, one at £25 and the other at £20.

In no case can a claim be made for a pension, which must be borne in mind in contrasting these figures with those of English asylums.

For the sake of comparison I add the following table from the last report of the Portsmouth Borough Asylum (England), where the patients number 450:—

OFFICERS.—Medical superintendent, £480 per annum, with unfurnished residence, light, fire, garden produce, milk, and washing; assistant medical officer, £120 per annum, with board, furnished apartments, gas, coal and washing; chaplain, £180 per annum, non-resident; clerk of the asylum and steward, £200 per annum, non-resident.

ATTENDANTS (male department).—One head attendant, £40 per annum, with board, lodging, washing and uniform; three night attendants, 18s. 6d. per week, with one meal per night, non-resident; one charge attendant £30 per annum, one ditto £27, two ditto £26 10s., one ditto £25, each with board, lodging, washing and uniform; one second class attendant £25, two ditto £23 10s., two ditto £23, each with board, lodging, washing and uniform; one third class attendant £23, one ditto £22, six ditto £21, each with board, lodging, washing and uniform; one hall porter £27, one ditto £19 5s., each with board, lodging, washing and uniform.

NURSES (female department).—One housekeeper and chief nurse, £55 per annum, with furnished apartments, board, washing, &c.; one organist, £30 per annum, with ditto; one needle mistress, £27 per annum, with board, lodging, washing and uniform; one night nurse, 14s. per week, with one meal per night, non-resident; one night nurse £21 per annum, one ditto £21 10s., each with board, lodging, washing and uniform; one charge nurse £24, four ditto £22, one ditto £21, one ditto £20, each with board, lodging, washing and uniform; two second class nurses, £24, two ditto £20, one ditto £19 10s., one ditto £19, each with board, lodging, washing and uniform; two third class nurses £17 10s., two ditto £17, one ditto £16 10s., six ditto £16, each with board, lodging, washing, and uniform.

It is stated in the report of the Kingston Asylum, that the value of the produce of the farm and garden amounted last year to upwards of £1,370. Two hundred and twenty three patients performed 57,244 days' work during the year. When I visited this asylum, a circumstance had just occurred which displays, in its after-history, a curious condition of Canadian law. A male patient escaped from the asylum and made a criminal assault, for which he was arrested and tried. Incredible as it may seem, the opinion of the medical superintendent of the asylum was never sought. The man was found guilty, and sentenced to six months' hard labour in jail without the question of his insanity being gone into. The Judge stated that he must be lenient under the circumstances, but what these were have not been stated. Having read the history of this case, I should regard him as a most dangerous lunatic, and should not be surprised if he commits some frightful crime when he regains his liberty. It is difficult to understand why he was not placed in the criminal asylum, where he would certainly have been prevented doing any injury to society.

I am informed that in the old Lunacy Act (prior to 1871) there was a clause which should not have been repealed, viz., the provision made for the detention of criminal lunatics in the criminal asylum as soon as their sentences expired. At present the asylum-authorities are forced to receive all criminal lunatics and insane criminals belonging to the province of Ontario at the time their sentences have expired. This state of affairs is, as might be expected, most unfortunate for the Kingston Asylum, for it is made the repository for all these criminals, and their influence is anything but salutary.

I append the dietary table at the Kingston Asylum for one week in July of this year :—

Days of Week.	Breakfast.	Dinner.	Tea.
Monday	Rice and milk. Coffee, bread and butter.	Barley soup. Beef, potatoes and bread.	Tea, bread and butter.
Tuesday	Porridge and milk. Coffee, bread and butter.	Roast beef, potatoes, and bread.	Cheese. Tea, bread and butter.
Wednesday	Cold meat. Coffee, bread and butter.	Barley soup. Beef, potatoes and bread.	Rhubarb. Tea, bread and butter.
Thursday	Porridge and milk. Coffee, bread and butter.	Plum pudding. Roast beef, potatoes and bread.	Tea, bread and butter.
Friday	Porridge and milk. Coffee, bread and butter.	Boiled fish. Beef, potatoes and bread.	Buns. Tea, bread and butter.
Saturday	Porridge and milk. Coffee, bread and butter.	Pea soup. Pork, beef, potatoes and bread.	Tea, bread and butter.
Sunday	Coffee, bread and butter.	Beans. Roast beef and bread.	Rhubarb. Tea, bread and butter.

With this dietary may be compared that of the Portsmouth Borough Asylum (England), which is probably above the average dietary of County and Borough Asylums :—

BREAKFAST (daily).

Males.—8ozs. bread, ½oz. butter, 1 pint tea, coffee or cocoa.

Females.—6ozs. bread, ½oz. butter, 1 pint tea, coffee or cocoa.

SUPPER (daily).

Males.—8ozs. bread, 2ozs. cheese or ½oz. butter; 1 pint tea.

Females.—6ozs. bread, 2ozs. cheese or ½oz. butter; 1 pint tea.

DINNER.

Sunday.—16 to 18ozs. suet pudding, with treacle sauce, and the addition of fruit in the summer and dried fruit in the winter—males and females. 3ozs. of meat where ordered.

Monday.—5ozs. meat males, and 4ozs. females; vegetables not less than 1lb.

Tuesday.—3ozs. tinned meat, males and females; vegetables as on Monday.

Wednesday.—2 pints soup, 2ozs. meat, 5 ozs. bread—males. 1½ pint soup, 4ozs. bread—females.

Thursday.—Meat pie, 12ozs.—males; 10ozs.—females. ½lb. potatoes or ¼lb of other vegetables.

Friday.—Same as Monday.

Saturday.—1lb. fish, males and females; vegetables same as Monday.

Half-pint of ale daily for dinner, except on Wednesday, for both males and females.

Women working in the laundry have bread and cheese and half-pint of ale for lunch, with meat and ale for dinner on Wednesday; also extra tea at 3 p.m.

Women scrubbing in wards have bread and cheese daily for lunch, with half-pint of ale. Men the same.

Men working in the shops or on the farm have half-pint of ale and bread and cheese at 10 a.m., and ale at four o'clock.

Meat pie contains 3ozs. of meat without bone for each patient. Soup is made from liquor of boiled meat, thickened with pearl barley, &c., to which are added vegetables, herbs, &c.

From the asylum I proceeded to visit the *Penitentiary*, which is in the vicinity, accompanied by Dr. Clarke. Mr. Creighton, the warder, who showed me over, is a very kindly gentleman. The prison appears to be in excellent order. There is a separate modern building for 43 criminal lunatics. The number on the day of my visit was 37. The character of the cells is, I am sorry to say, similar to that of a prison, and, so far as I could judge, the patients are treated with almost as much rigour as convicts, though not dressed in prison garb. This is wrong. Either they are or are not lunatics. If they are, they ought to be very differently cared for, while every precaution to prevent escape is taken. In the basement are "dungeons," to which patients when they are refractory are consigned as a punishment, although the cells above are in all conscience sufficiently prison-like. The floors of the cells are of stone, and would be felt to be a punishment by any patient in the asylums of Ontario.

In a day-room above the ground floor a number of patients were congregated, moody and apathetic. Some were in mechanical restraint.

Two men in the cells had once been patients in the asylum. One, with whom we conversed at the iron gate of his dungeon, laboured under a distinct delusion of there being a conspiracy against him. It was certainly not very likely to be dispelled by the dismal stone-floor dungeon in which he was immured, without a seat, unless he chose to use the bucket intended for other purposes, which was the only piece of furniture in the room. Surely something will be done to terminate a condition of such unnecessary hardship. For criminals of the worst class, this building is no doubt admirably suited, but it is astonishing that it should have been constructed for lunatics in recent times. In these remarks no reflection is for a moment cast on the excellent warder of the Penitentiary. As to what the Visiting Medical Officer does by way of medical treatment of these patients, or to secure their comfort, I shall not attempt to give an opinion.

I hasten to remark that the Penitentiary is not under the control of the Province, but the Dominion; otherwise, judging from the asylums of Ontario, it would, I have no doubt, be in a totally different condition.

It will thus be seen that the Province of Ontario possesses in its asylums excellent institutions, in which modern views and the results of experience in other countries are vigorously and intelligently applied; in which employment is being carried out more and more to the extent consistent with the comfort of the patients; in which mechanical restraint is not resorted to unless all other means have failed, and in which a good example of segregation is exhibited, the usually constructed asylum being supplemented by an annexe or cottages adapted for particular classes. That such a system as this works well, no one who has seen it in operation in British or other asylums will be surprised to hear.

To the preceding notices of the asylums in the provinces of Quebec and Ontario a few additional notes are appended.

New Brunswick.

New Brunswick, which had in 1881 a population of 321,129, has one asylum in the Province, for the insane, the property of the Local Government.

This institution is controlled and inspected by eight Commissioners. Dr. J. T. Steeves is the medical superintendent, and there is one assistant medical officer. The asylum is located at Saint John, on a height overlooking the Saint John river. It was opened in 1848, and has been enlarged since that date. The last report states that the number of patients is 380, of whom 198 were males and 182 were females, but I understand that the capacity of the building should allow of only 325 patients. If so, the Saint John Asylum is no exception to the usual over-crowding, and fails to meet the requirements of the Province. I am glad to hear that Dr. Steeves is agitating for the erection of cottages, and proposes to place them on a large farm at a distance from the present institution.

Dr. Clarke, the assistant medical officer at the Kingston Asylum, to whom I am largely indebted for the information contained in these supplementary notes, informs me that he visited this asylum in July (1884) and found it well ordered and managed. The wards were quite as cheerful and at least as homelike as in the institutions of Ontario. Great attention was given to decorations; pictures and flowers being placed in most parts of the building. The patients were cheerful, clean, and evidently well-cared for. Mechanical restraint was reduced to a minimum, and employment provided for many of the patients, although not to the extent carried out in Ontario. The farm is a very small one, comprising but a few acres, so that it is difficult for the superintendent to employ the men as he would wish. The cost of maintenance last year was £26 per head.

I observe that, in the last annual report, the nativity of the patients admitted since 1875 is given, showing that out of 1,249, 884 were from New Brunswick, and 57 from other provinces of Canada, while of the remaining non-Canadian population (308), no less than 220 were of Irish extraction.

Of the 380 patients in the asylum, only 47 were regarded as curable.

Nova Scotia.

There is one provincial asylum or hospital for the insane of Nova Scotia with its population of 440,585. It was opened in 1859. It is situated at Halifax, is the property of the Local Government, and is under the inspection and control of a Board of Commissioners of Public Charities, six in number. Dr. Alexander P. Reid is the medical superintendent, and there is one assistant physician. The number of patients in the asylum, Jan. 1, 1884, was 400; 193 men and 207 women. The weekly rate for maintenance was 10s. 10d. The asylum is said to be well managed and not behind the times. How far it fails to accommodate the number of insane in the Province I am unable to state.

The statistical tables give the information so rarely vouchsafed to the inquirer into the recoveries and the deaths of patients since the opening of the hospital, along with the admissions and the average number resident. From this it appears that the recoveries amounted to 44·6 per cent. of the admissions, while the deaths, calculated upon the average number resident, amounted to 6 per cent. It may be observed that the Medico-Psychological Association's tables are adopted at this asylum, although the recent amendments have not been introduced.

Prince Edward's Island.

The provincial asylum is situated at Charlottetown. The population of the Province in 1881 was 108,928. Dr. E. S. Blanchard is the medical superintendent. The asylum is under the management and inspection of a Board of Trustees. The institution has accommodation for 112 patients. It is stated that the number of insane known to exist in the Province, but not in the asylum, is 65. The cost per patient in 1884 was about £35, including not only ordinary maintenance, but repairs, salaries, and wages. Many important changes have been made of late in the way of improvement.

Manitoba.

As might be expected, this Province has had until recently no proper accommodation for its insane, the number of whom must be considerable in a Province, the population of which has so vastly increased since the time of the last census (1881), when it was 50,000. In the absence of an asylum, lunatics have been kept in the penitentiary or sent to Ontario. An asylum is now being built at West Selkirk, which, when completed, will be large enough to meet the requirements of the Province for some time to come. The Local Government have taken temporary quarters in place of the penitentiary at Lower

Fort Garry. Dr. Young is appointed medical superintendent at the new asylum. It is understood that the Manitoba Lunacy Law resembles that of Ontario.

I am unable to state what provision has been made for the insane of the Province of British Columbia.

Newfoundland.

The asylum is at St. John's. It accommodates about 150 patients, and is superintended by Dr. Stabb.

APPENDICES.

APPENDICES A. AND B.

TABLE I.—General Statement of Insane in Hospitals and Lunatic Asylums in the United States and British Provinces on September 30, 1883.

Extracted from the First Report of the Committee on Lunacy of Board of Public Charities of the State of Pennsylvania, September 30, 1883.

I am unable to explain how it is that the capacity of hospitals and asylums for the insane is represented as greater than the number of patients, seeing that even in 1880 the overcrowding amounted to several thousand patients.

The number of insane given in this Table as being in asylums in British North America is less by one thousand than the number reported in Table II. in consequence of the Committee on Lunacy having omitted the Montreal Asylum, which I have added to the latter Table. With this figure the aggregate number would be 57,685, instead of 56,685.

TABLE II.—In the above-mentioned Report it is stated that "several institutions of the United States in this Table, being departments of Almshouses, might with propriety have been excluded, but as they had been included in a list of institutions for the insane published in the compendium of the Census of 1880, we thought best to retain them, in order that a comparison could be made with their condition at that date, and one very nearly corresponding thereto three years later" (p. 7). They number 147.

TABLE I.

		UNITED STATES.			BRITISH NORTH AMERICA.			AGGREGATE.		
		Male.	Female.	Total.	Male.	Female.	Total.	Male.	Female.	Total.
Capacity of Institutions		26,466	25,447	51,913	2,397	2,375	4,772	28,863	27,822	56,685
Total of Patients		26,248	25,567	51,815	2,288	2,235	4,523	28,536	27,802	56,338
Race	White	24,734	24,089	48,823	2,273	2,213	4,486	27,007	26,302	53,309
	Coloured	1,514	1,478	2,992	15	22	37	1,529	1,500	3,029
Nativity	Native	16,419	15,467	31,886	1,575	1,480	3,055	17,994	16,947	34,941
	Foreign	9,829	10,100	19,929	713	755	1,468	10,542	10,855	21,397
Class	Private	2,401	2,831	5,232	94	141	235	2,495	2,972	5,467
	Public	23,847	22,736	46,583	2,194	2,094	4,288	26,041	24,830	50,871
Complications	Epileptic	1,865	1,311	3,176	153	110	263	2,018	1,421	3,439
	Paralytic	897	289	1,186	40	23	63	937	212	1,249
	Homicidal	1,563	884	2,447	35	20	55	1,598	904	2,502
	Suicidal	1,093	1,048	2,141	81	63	144	1,174	1,111	2,285
Insane Convicts		482	41	523	21	15	36	503	56	559
Criminal Insane		491	89	560	50	12	62	541	101	642

APPENDIX B.

TABLE II.—Insane in Hospitals and Lunatic Asylums in the United States and British Provinces, September 30, 1883.

Extracted from the First Report of the Committee on Lunacy of Board of Public Charities of the State of Pennsylvania, September 30, 1883.

The number of insane reported to be in these institutions at this date is very considerably larger than at the time of the Census, 1880.

[Explanation of signs: * State or territorial asylums. † Catholic Institutions. ‡ Owned by the United States.]

NAME OF INSTITUTION.	LOCATION.	NUMBER.	EXECUTIVE OFFICER.
Alabama Insane Hospital	Tuskaloosa	497	Peter Bryce, M.D.
Arkansas.			
*Arkansas State Lunatic Asylum	Little Rock	247	C. C. Forbes, M.D.
California.			
*Napa State Asylum for the Insane	Napa City	1,277	E. T. Wilkins, M.D.
*Insane Asylum of the State of California	Stockton	1,199	W. T. Browne, M.D.
Pacific Asylum	Stockton	53	Asa Clark, M.D.
Colorado.			
*Colorado State Lunatic Asylum	Pueblo	79	P. R. Thombs, M.D.
Connecticut,			
*Connecticut Hospital for Insane	Middletown	861	Abram Marvin Shew, M.D.
Retreat for the Insane	Hartford	138	Henry G. Stearns, M.D.
Home for Nervous Invalids	Litchfield	20	Henry W. Buel, M.D.
Cromwell Hall	Cromwell	18	W. B. Hallock, M.D.
Home for Mild Forms of Insanity, Inebriety, and the Opium Habit	Stamford	—	George F. Foote, M.D.
Dakota.			
*Dakota Hospital for the Insane	Yankton	106	S. B. McGlumphy, M.D.
District of Columbia.			
‡Government Hospital for the Insane	Washington	1,008	W. W. Godding, M.D.

TABLE II.—Continued.

NAME OF INSTITUTION.	LOCATION.	NUMBER.	EXECUTIVE OFFICER.
Florida.			
*Florida Insane Asylum	Chattachoochee	128	J. H. Randolph, M.D.
Georgia,			
*Lunatic Asylum, State of Georgia	Milledgeville	1,131	T. O. Powell, M.D.
Illinois.			
*Illinois Northern Hospital for the Insane	Elgin	525	Edwin A. Kilbourne, M.D.
*Illinois Eastern Hospital for the Insane	Kankakee	488	Richard B. Dewey, M.D.
*Illinois Central Hospital for the Insane	Jacksonville	627	H. F. Carriel, M.D.
*Illinois Southern Hospital for Insane	Anna	534	Horace Wardner, M.D.
Cook County Hospital for Insane	Jefferson	534	F. C. Spray. M.D.
Bellevue Place	Batavia	28	R. J. Patterson, M.D.
McFarland Retreat for the Insane	Jacksonville	32	Andrew McFarland, M.D., LL.D.
Indiana.			
*Indiana Hospital for the Insane	Indianapolis	1,090	W. B. Fletcher, M.D.
Iowa.			
*Iowa Hospital for the Insane at Mt. Pleasant	Mt. Pleasant	513	H. A. Gilman, M.D.
*Iowa Hospital for the Insane at Independence	Independence	564	Gersham H. Hill, M.D.
Kansas.			
*Kansas State Insane Asylum	Osawatomie	418	A. H. Knapp, M.D.
*Kansas State Insane Asylum	Topeka	238	A. G. Tenney, M.D.
Kentucky.			
*Central Kentucky Lunatic Asylum	Anchorage	353	R. H. Gale, M.D.
*Western Kentucky Lunatic Asylum	Hopkinsville	516	James Rodman, M.D.
*Eastern Kentucky Lunatic Asylum	Lexington	644	R. O. Chenault, M.D.
Louisiana,			
*Insane Asylum of Louisiana	Jackson	508	J. Welsh Jones, M.D.
City Insane Asylum	New Orleans	—	
†Louisiana Retreat	New Orleans	117	E. T. Shepard, M.D.
Maine.			
*Maine Insane Hospital	Augusta	462	Biglow T. Sanborn, M.D.
Maryland.			
*Maryland Hospital for the Insane (Spring Grove)	Cantonsville	404	Richard Gundry, M.D.
Bay View Asylum (Department of City Alms-house)	Baltimore	211	Charles Carroll.

Maryland.—Continued.				
†Mount Hope Retreat	...	Baltimore	470	William H. Stokes, M.D.
Mattey Hill Sanitarium	...	St. Denis	24	I. B. Conrad, M.D.
Haarlem Lodge	...	Cantonsville	10	William F. Stewart, M.D.
Massachusetts.				
*Worcester Lunatic Hospital	...	Worcester	731	John G. Park, M.D.
*Taunton Lunatic Hospital	...	Taunton	633	J. P. Brown, M.D.
*Northampton Lunatic Hospital	...	Northampton	469	Pliny Earle, M.D.
*Danvers Lunatic Hospital	...	Danvers	721	Wm. B. Stealman, M.D.
*Asylum for Chronic Insane	...	Worcester	392	Hosea M. Quinby, M.D.
*Insane Asylum of State Alms-House	...	Tewkesbury	252	O. Irving Fisher, M.D.
Boston Lunatic Hospital	...	South Boston	187	Theodore W. Fisher, M.D.
McLean Asylum for the Insane	...	Somerville	164	Edward Cowles, M.D.
Essex County Receptacle for Insane	...	Ipswich	62	Y. G. Hurd.
Private Hospital	...	Brookline	5	Walter Channing, M.D.
Adam Nervine Asylum	...	Jamaica Plain	31	F. W. Page, M.D.
Shady Lawn	...	Northampton	11	Austin White Thompson, M.D., A.M.
Cutter Retreat	...	Pepperell	2	J. S. N. Howe,
Family Home	...	Highlands, Winchendon	10	Ira Russell, M.D.
Herbert Hall	...	Worcester	16	Merrick Benis, M.D.
Michigan.				
*Michigan Asylum for the Insane	...	Kalamazoo	752	George C. Palmer, M.D.
*Eastern Michigan Asylum	...	Pontiac	640	Henry M. Hurd, M.D.
†Michigan State Retreat	...	Detroit	95	J. G. Johnson, M.D.
Minnesota,				
*Minnesota Hospital for Insane	...	St. Peter	690	Cyrus K. Bartlett, M.D.
*Second Minnesota Hospital for Insane	...	Rochester	327	J. E. Bowers, M.D.
Mississippi.				
*Mississippi State Lunatic Asylum	...	Jackson	455	T. J. Mitchell, M.D.
Missouri.				
*Missouri State Lunatic Asylum	...	Fulton	519	T. R. H. Smith, M.D.
*Missouri State Lunatic Asylum, No. 2	...	St. Joseph	268	George C. Catlett, M D.
St. Louis Insane Asylum	...	St. Louis	495	Charles W. Stevens, M.D
†St. Vincent's Institution (for the Insane)	...	St. Louis	177	{ J. K. Baudoy, M.D. A. B. Shaw, M.D
†Alexian Brothers Misericordia Insane Asylum	...	St. Louis	20	Bro. Jodneus Schiffer
City Poor-House—Insane Department	...	St. Louis	403	William J. Beck.

Table II.—Continued.

Name of Institution.	Location.	Number.	Executive Officer.
Montana Territory.			
Warm Springs Insane Asylum	Deer Lodge City	69	{ A. H. Mitchell, M.D. Charles F. Masslgbrod, M.D.
Nebraska.			
*Nebraska State Hospital for Insane	Lincoln	266	H. P. Matthewson, M.D.
Nevada.			
*Nevada State Insane Asylum	Reno	146	Simeon Bishop, M.D.
New Hampshire.			
*New Hampshire Asylum for the Insane	Concord	296	Charles P. Bancroft, M.D.
New Jersey.			
*New Jersey State Lunatic Asylum	Trenton	628	John W. Ward, M.D.
*State Asylum for the Insane at Morristown	Morris Plains	707	H. A. Buttolph, M.D.
Burlington County Insane Asylum	Pemberton	67	Theodore B. Gaskill.
Camden County Asylum	Blackwood	77	Mrs. Adaline B. Stiles.
Essex County Asylum for Insane	Newark	356	John Leonard.
Hudson County Lunatic Asylum	Jersey City	216	George W. King, M.D.
Passaic County Lunatic Asylum	Paterson	42	Cornelius L. Petry.
New York.			
*New York State Lunatic Asylum	Utica	604	John P. Gray, M.D., LL.D.
*Hudson River State Hospital	Poughkeepsie	308	Joseph M. Cleveland, M.D.
*Buffalo State Asylum for the Insane	Buffalo	329	Judson B. Andrews, M.D.
*New York State Homœopathic Asylum for Insane	Middletown	260	Selden H. Talcott, A.M., M.D., Ph.D.
*Willard Asylum for Insane	Willard	1,758	P. M. Wise, M.D. (1885)
*Binghampton Asylum for the Chronic Insane	Binghampton	425	T. S. Armstrong, M.D.
State Asylum for Insane Criminals	Auburn	148	Carlos F. MacDonald, M.D.
*State Emigrant Insane Asylum	Ward's Island	108	M. R. C. Peck, M.D.
New York City Lunatic Asylum	Blackwell's Island	1,461	Thomas M. Franklin, M.D.
Branch New York City Asylum for the Insane	Randall's Island	128	James R. Healy, M.D.
New York City Branch Lunatic Asylum	Hart's Island	433	Andrew Egan, M.D.
New York City Asylum for the Insane	Ward's Island	1,357	A. E. Macdonald, M.D.
Homœopathic Hospital	Ward's Island	—	T. M. Strong, M.D.

New York—Continued.			
King's County Insane Asylum	Flatbush, Long Island	831	} John C. Shaw, M.D.
King's County Hospital for Incurables		376	Frederick Busch.
Erie County Alms-house (Insane Department)	Buffalo Plains	269	M. L. Lord, M.D.
Monroe County Insane Asylum	Rochester	238	A. A. Aldrich, M.D.
Onondaga County Insane Asylum	Onondaga	113	David Rogers, M.D.
Queen's County Insane Asylum	Mineola, Long Island	126	J. D. Lomax, M.D.
Rensselaer County Lunatic Asylum	Troy	91	Charles H. Nichols, M.D.
Bloomingdale Asylum for Insane (Dept. of N. Y. Hos.)	New York	225	H. Earnest Schmid, M.D.
† St. Vincent's Retreat for the Insane	Harrison	39	Lydia Kleth.
Keith Home	Brooklyn	10	
Private Hospital	Pleasantville	9	George C. S. Choate, M.D.
Sanford Hall	Flushing, Long Island	30	J. W. Barstow, M.D.
Brigham Hall	Canandaigua	56	D. R. Burrell, M.D.
† Providence Lunatic Asylum	Buffalo	120	Sister Rosaline Brown.
North Carolina.			
*North Carolina Insane Asylum	Raleigh	250	Eugene Grissom, M.D.
*Eastern N. C. Insane Asylum (coloured insane)	Goldsboro'	109	J. D. Roberts, M.D.
*Western North Carolina Insane Asylum	Morganton	152	P. L. Murphy, M.D.
Ohio.			
*Columbus Asylum for Insane	Columbus	855	H. C. Rutter, M.D.
*Cleveland Asylum for the Insane	Cleveland	627	Jamin Strong, M.D.
*Dayton Asylum for the Insane	Dayton	602	H. A. Toby, M.D.
*Athens Asylum for the Insane	Athens	618	A. B. Richardson, M.D.
North-Western Ohio Hospital for Insane	Toledo	113	J. J. Lawless, M.D.
Longview Asylum	Carthage	674	C. A. Miller, M.D.
Cincinnati Sanitarium	College Hill	54	Orpheus Everetts, M.D.
Oxford Retreat	Oxford	23	D. A. Morse, M.D.
Oregon.			
*Oregon State Insane Asylum	Salem	364	Horace Carpenter, M.D.
*Oregon Hospital for Insane	East Portland	—	B. E. Josephi, M.D.
Pennsylvania.			
*Pennsylvania State Lunatic Hospital	Harrisburg	398	J. Z. Gerhard, M.D.
*State Hospital for the Insane	Danville	327	S. S. Schultz, M.D.
*State Hospital for the Insane	Norristown	1,005	Robert H. Chase, M.D.
			Alice Bennett, M.D.
State Hospital for the Insane	Warren	423	John Ourwen, M.D.
Asylum for the Relief of Persons deprived of the use of their Reason	Frankford, Philadelphia	97	John C. Hall, M.D.
Insane Department of the Philadelphia Hospital	Philadelphia	617	D. D. Richardson, M.D.
Pennsylvania Hospital for the Insane	Philadelphia	373	John B. Chapin, M.D. (1884).

TABLE II.—Continued.

NAME OF INSTITUTION.	LOCATION.	NUMBER.	EXECUTIVE OFFICERS.
Pennsylvania.—Continued.			
Western Pennsylvania Hospital for the Insane	Dixmont	498	Joseph A. Reed, M.D.
Burn Brae	Clifton Heights	31	Robert A. Given, M.D.
Private House	Wawa	4	Alfred T. Livingston, M.D.
Insane Department, Lancaster County Hospital	Lancaster	100	John H. MacCreary, M.D.
Rhode Island.			
Butler Hospital for the Insane	Providence	192	John W. Sawyer, M.D.
*State Insane Asylum	Cranston	290	Charles H. Hunt.
South Carolina.			
*South Carolina Lunatic Asylum	Columbia	613	Peter E. Griffin, M.D.
City Hospital, Lunatic Department	Charleston	12	Prof. J. C. Chazal, M.D.
Tennessee.			
*Tennessee Hospital for the Insane	Nashville	404	John H. Callender, M.D.
Texas.			
*Texas State Lunatic Asylum	Austin	410	A. N. Denton, M.D.
Utah.			
Salt Lake City Insane Asylum	Salt Lake City	25	S. B. Young, M.D.
Insane House (County Jail)	Saint George	2	A. P. Hardy.
Vermont.			
Vermont Asylum for the Insane	Brattleboro	427	Joseph Draper, M.D.
Virginia.			
*Eastern Lunatic Asylum	Williamsburg	447	Richard A. Wise, M.D.
*Western Lunatic Asylum	Staunton	534	R. S. Hamilton, M.D.
*Central Lunatic Asylum	Richmond	409	David F. May, M.D.
Pinel Hospital	Richmond	—	James D. Moncure, M.D.
Washington Territory.			
Hospital for the Insane	Fort Stellacoom	133	John W. Waughop, M.D.

West Virginia.			
*West Virginia Hospital for Insane	Weston	662	William J. Bland, M.D.
Wisconsin.			
*Wisconsin State Hospital for the Insane	Madison	469	S. B. Buckmaster, M.D. (1884).
*Northern Hospital for Insane	Winnebago	596	R. M. Wigginton, M.D. (1884).
Milwaukee Asylum for Insane	Wauwatosa	318	— Scrivener, M.D. (1884).
Sheboygan County Chronic Insane Asylum	Sheboygan	49	A. J. Wiffin.
		51,815	
British Provinces.			
Prince Edward's Island Hospital for Insane	Charlottetown, P. E. Island	108	Edward S. Blanchard, M.D.
Nova Scotia Hospital for Insane	Halifax, N.S.	399	A. P. Reed, M.D.
Asylum for the Insane	Kingston, Ont.	449	W. G. Metcalf, M.D.
Asylum for the Insane	Toronto, Ont.	703	Daniel Clark, M.D.
Asylum for the Insane	London, Ont.	884	R. M. Bucke, M.D.
Asylum for the Insane	Hamilton, Ont.	547	James M. Wallace, M.D.
Provincial Lunatic Asylum (Lancaster)	St. John, N.B.	371	James T. Steeves, M.D.
Quebec Lunatic Asylum	Quebec, Canada	908	{ J. E. J. Landry, M.D. { F. E. Roy, M.D.
†Longue Pointe Asylum	Montreal, Canada	1,000	Ste Thérèse.
St. John's Lunatic Asylum	St. John's, Newfoundland	154	Henry H. Stabb, M.D.
Homewood Retreat	Guelph, Ont.	...	Stephen Lett, M.D.
Asylum for the Insane	West Selkirk, Manitoba	Not open	— Young, M.D. (1885).
		5,523	

APPENDIX C.

TABLE III.—Showing Total Number of Insane in each of the United States, and the Proportion to the Population in 1880. Prepared by Prof. A. O. Wright, Madison.

It must always be borne in mind that under "Insane" the American Census does not include idiots. If these be added, the proportion to the population would be raised from 1 in 545 to 1 in 297.

The insane, according to the last Census, exceed the idiots by about 19 per cent., but the Census Office believes this to be too high, and maintains that the number of idiots "must be nearly or quite equal to that of the insane"—an opinion which we cannot endorse if the experience of Great Britain may be applied to the United States. The test of idiocy adopted by the Census Office is the age of puberty, all returns of mental affection under 12 years of age in girls, and 14 in boys, being classed under idiocy. It cannot be said that this or any other arbitrary test which might be adopted is satisfactory. The distinction, therefore, between these two classes must not be taken as at all accurate. In fact, it is difficult to credit the result that the number of idiotic males in the United States exceeds the insane males, as appears from the following figures in the census 1880:—

CLASS.	MALES.	FEMALES.	TOTAL.
Insane	44,391	47,568	91,959
Idiots	45,309	31,586	76,895
	89,700	79,154	168,854

Table III.

States.	Proportion of Insane.	Total Insane.
Maine	1 in 421	1,542
New Hampshire	328	1,056
Vermont	327	1,015
Massachusetts	343	5,127
Connecticut	361	1,723
Rhode Island	404	684
New York	360	14,111
New Jersey	470	2,405
Pennsylvania	516	8,304
Ohio	439	6,286
Indiana	560	3,530
Illinois	599	5,134
Michigan	586	2,796
Wisconsin	520	2,526
Iowa	638	2,544
Minnesota	682	1,045
Delaware	721	198
Maryland	504	1,857
Virginia	627	2,411
West Virginia	629	982
North Carolina	690	2,028
South Carolina	895	1,112
Georgia	907	1,697
Kentucky	591	2,789
Tennessee	517	2,404
Missouri	636	3,310
Florida	1,054	253
Alabama	830	1,521
Mississippi	999	1,147
Louisiana	943	1,602
Arkansas	1,017	789
Texas	1,021	1,554
Nebraska	1,006	450
Kansas	996	1,000
Dakota	2,037	72
Colorado	1,966	99
Montana	663	59
Wyoming	5,187	4
Utah	953	154
New Mexico	781	152
California	316	2,503
Oregon	462	378
Washington Territory	556	135
Nevada	2,008	31
Idaho	2,038	16
District of Columbia	189	938
Total average United States	1 in 545	
Total insane		91,997

Note.—These figures show a lower rate of insanity in the more recently settled States.

APPENDIX D.

TABLE IV.—Showing amount of Restraint and Seclusion in twenty Hospitals for the Insane in 1882.

This Table is extracted from a pamphlet entitled "On Restraint and Seclusion in American Institutions for the Insane," by H. M. Bannister and H. N. Moyer, 1882; reprinted from the "Journal of Nervous and Mental Disease," Vol. ix., No. 3. Dr. Bannister is the senior assistant medical officer at the Kankakee Asylum, and has been an able contributor to the above Journal.

The observation I have made on the returns of restraint given in 1880 may be repeated here—that the amount was much greater then than now. In a letter received from Dr. Chapin, July, 1885, he says: "A great change for the better has taken place in hospitals on this side the Atlantic during the past five years. At the last annual meeting of the Conference of Charities and Corrections, it was announced that in 12 asylums of this country mechanical restraint was practically abolished."

In studying this Table, the following remarks by the physicians who prepared it, should be read :—

"In No. 3, mittens are not reported as restraint, but are used to a slight extent. If reported, as in some of the other institutions, they would slightly increase the figures of restraint.

No. 5 suffers to a very marked extent from the evils of overcrowding and insufficient and incomplete accommodations, and has no properly constructed 'back wards' on the female side.

The same is the case with No. 10.

In the one instance of restraint reported at No. 9, it was ordered by the assistant physician and, in the opinion of the superintendent, by an 'error of judgment,' it not being really needed.

In No. 10, the only form of restraint used during the month was the crib-bed at night.

In No. 11, the restraint and seclusion were, with only a few exceptions, used in the female department.

The report from No. 12 was made up for the month of January instead of February.

The superintendent of No. 14 writes that the total of hours of restraint and seclusion is somewhat in excess, as a number of short intermissions of from half an hour to an hour are not noted in the daily ward reports.

In No. 15, the sleeping doses at night are evidently included in the 'chemical restraint' returns, instead of being reported separately.

Nos. 17 and 18 were not reported on the blanks furnished by us, but our columns were filled from other reliable figures supplied us by the authorities of these hospitals.

In No. 20, the crib-bed is not reported as restraint, though we believe it is used. The restraint reported was, in a great majority of cases, applied only at night.

Several of the physicians that have reported the use of 'chemical restraint' in the hospitals under their charge, also send explanatory notes stating that the term is hardly applicable, as it is impossible for them to separate entirely the use of sedatives to control excitement from the legitimate therapeutics of insanity. As one of them states it, 'the primary object of the exhibition of this class of remedies is physiological and curative; the secondary, or that of restraint, being only incidental.' There is considerable force in this view, and we do not attach the same value to the returns in these columns as to the others in the table. Moreover, it must be remembered that medicine and other therapeutic measures have a legitimate use as palliatives when there is absolutely no hope of any curative effect. Thus we give bromides to certain insane epileptics merely to modify the manifestations of the disease, which is itself past curing, and this in the strict sense of the term might be called chemical restraint. If all such instances were included under this head, there is little doubt but that 'chemical restraint' might have been reported from every hospital on our list.

After allowance is made for all the possible incompleteness and lack of uniformity of the table, it is still sufficiently suggestive, and affords ample opportunity for discussion. The first point it suggests to the examiner is that there is a wide range of practice in hospitals on this side of the Atlantic as regards the use of restraint, seclusion, and sedative medication. If we even go to the extent of throwing out all the figures of 'chemical restraint,' as not sufficiently separated from the ordinary therapeutic treatment and night sedatives to be altogether reliable, we find the percentage of patients nightly taking sleeping draughts ranges in the different institutions from nothing up to over eleven per cent. of the population; that the average daily percentage of patients restrained rises in the same manner to over six per cent.; and that of seclusion to nearly three per cent. of the whole number of patients. The average percentage of restraint, according to these figures, would be about one and a half, and this cannot, in our opinion, be taken as too high an estimate of the actual average percentage in the asylums of this country. Our list includes several institutions that are practically 'non-restraint' and exceptional in that respect, and any figures derived from this table, including these institutions, must be affected by this fact. We have reason to believe that the institutions that did not report to us would show fully as high a ratio of restraint as the majority of those from which we heard, and that while the percentages of Nos. 12, 16, and 17 are almost certainly above the average, those of Nos. 3, 4, 7, 9, and 19 are more than as much below it. It is very possible, indeed, highly probable, that there are some asylums in this country that would show higher figures of restraint than any from which we have had returns.

The English Commissioners' report for 1880 shows that there is vastly less seclusion in England than in this country, the special minutes of inspection of sixty-two public asylums, with a population of about 40,000, showing only 3,240 instances of seclusion in them for a period averaging, in the whole number, over a year; whereas nineteen American institutions, containing less than eleven thousand patients, had in a single month 1,977 instances, or more than half what all the English asylums had in a whole year. In other words, seclusion, according to this ratio, would appear to be some twenty-eight or twenty-nine times as frequent on this side of the Atlantic as it is in England.

Space does not permit us to carry out fully the comparison of the statistics of restraint in the two sexes, as was our intention when we sent out our circulars and blanks, but some general data can be given. There were reported to us 1,655 occasions of the employment of restraint with males, lasting altogether $11,848\frac{1}{4}$ hours. The number of individuals reported as restrained was 79, of whom 12 were kept constantly in restraint, each day counting as one occasion; 79 males were also secluded, 6 of them constantly, on 847 occasions of altogether 6,622 hours.

Ninety-three females were reported as restrained for a total of 2,023 occasions and $13,984\frac{1}{4}$ hours, three of them being kept constantly restrained. Ninety females were secluded on 885 occasions, for $6,240\frac{3}{4}$ hours, two of them in constant seclusion. These figures, though of course incomplete, even for the institutions in our table, indicate that what is the common impression is correct—that insane women are more irritable and mischievous but less dangerous or violent than male patients, and that while, as a class, they seem to undergo restraint or seclusion more frequently, there are not so many of them that are so troublesome or dangerous as to be allowed no liberty at all."

TABLE IV.

Number of Institution	Daily average number of inmates during the month	Average daily number taking sedatives as "Chemical restraint"	Average number taking sleeping doses at night	Average daily number restrained	Whole number of instances of restraint	Total number of hours in which restraint was used	Average number of hours in each instance of restraint	Number of individuals restrained	Number of individuals in constant restraint	Percentage of restraint on whole number of patients. Daily average	Destruction	Discipline	Surgical	Masturbation	Violence to self	Violence to others	Maniacal excitement	Wristlets or muff	Camisole	Restrained to seat	Restrained to bed	Mittens	Crib Nights	Crib Days	Average daily number secluded	Whole number of instances of seclusion	Average number of hours in each instance of seclusion	Total number of hours of seclusion	Number of individuals secluded	Number of individuals in constant seclusion	Percentage of seclusion in whole number of patients (Daily average)
1	459	—	3.8	5.	139	2550	18.3	9	2	1.69	35	—	28	56	69	76	66	28	85	29	1	—	—	—	3.7	105	7.9	832	6	—	0.82
2	944	—	11.3	25.6	718	—	—	33	2	2.7	112	5	20	—	4	387	—	294	375	3	22	18	—	—	13.5	375	—	137	17	3	1.43
3	878	—	1.4	1.3	36	219	6.3	5	—	0.14	—	—	4	—	—	32	1	5	—	—	1	—	—	—	0.14	7	1.9	65	4	—	0.02
4	666	0.35	0.8	—	—	—	—	—	—	0.12	—	—	—	—	—	—	—	—	—	—	—	—	—	—	0.33	9	7.2	65	5	—	0.05
5	540	0.78	2.3	11.1	241	296	12.3	35	—	2.05	104	—	4	—	25	27	62	206	17	30	73	18	—	13	1.57	44	7.2	288	27	—	0.24
6	643	0.17	5.67	7.	312	4339	13.9	24	4	1.08	69	—	25	2	—	70	25	103	27	4	18	—	—	—	1.23	35	6.5	300	6	—	0.12
7	8.5	—	12.5	13.2	196	2235	11.4	—	—	0.	—	—	—	—	—	—	—	—	47	78	36	23	28	—	1.6	28	10.1	355	5	1	0.27
8	470	4.	5.67	0.005	370	6349	17.1	33	—	2.8	88	—	82	16	—	172	45	204	—	13	—	—	—	—	0.78	45	10.7	446	11	2	0.41
9	2.42	—	1.96	2.56	1	5	9.	1	—	0.1	—	—	1	—	28	—	—	—	69	—	—	—	—	—	12.17	341	9.9	3390	20	2	2.80
10	387	—	1.28	5.9	72	648	8.9	11	—	0.84	28	—	—	—	59	—	16	166	—	9	—	3	28	—	4.75	133	12.	31	7	1	0.25
11	1075	2.42	1.	1.5	166	1629	1.9	3	—	1.35	73	—	25	—	—	211	59	1	—	31	—	—	72	—	1.	22	12.	133	3	—	2.14
12	465	—	7.6	28.	26	1484	11.5	2	—	6.14	—	—	—	—	—	—	—	—	23	—	3	336	—	12.17	8	14.4	96	16	1	1.02
13	434	9.	6.	2.39	42	300	19.9	4	1	0.21	—	—	—	—	—	—	26	1	—	—	—	—	—	0.29	85	13.71	1235	2	—	0.07
14	203	5.	5.2	1.39	358	837	15.9	19	4	0.73	151	—	—	—	112	11	1	93	133	33	56	—	1	—	5.25	147	14.4	2016	17	—	1.47
15	420	1.8	—	15.1	423	5712	—	6	—	1.92	135	70	—	56	—	—	14	99	169	—	—	—	10	26	0.64	—	13.71	—	—	—	—
16	287	0.3	—	—	—	—	—	—	—	5.24	—	—	—	—	—	—	—	—	—	—	—	—	—	—	—	—	—	—	—	—	—
17	1075	—	6.	35.	1008	—	—	43	—	3.34	185	—	—	—	30	61	10	—	90	—	—	252	10	—	0.65	24	9.9	—	—	—	0.16
18	530	2.	13.3	2.39	67	113	10.3	5	1	0.45	42	—	4	31	—	—	7	135	—	—	—	21	—	—	6.7	187	9.9	1866	17	1	0.88
19	756	3.42	21.2	8.39	235	2242	9.5	21	—	1.59	99	—	—	—	—	—	—	—	—	8	—	159	—	—	6.28	176	11.6	2055	28	3	1.19
20	527	—	—	—	—	—	—	—	—	—	—	—	—	—	—	—	—	—	—	—	—	—	—	—	—	—	—	—	15	—	—

APPENDIX E.

TABLE V.—Showing the Number of Idiots in each State according to the Census 1880, and their Location at the same date.

These figures, as well as many others in this book, are extracted from the Compendium of the "Tenth Census of the United States," 2 parts, 1880, for a copy of which I am indebted to the Census Office, Washington. I take this opportunity of specially thanking Mr. Wines (expert and special agent), and also Mr. W. H. Barstow, for their courtesy and assistance, when I visited Washington.

Alabama	2,223
Arizona	11
Arkansas	1,374
California	507
Colorado	77
Connecticut	817
Dakota	80
Delaware	269
District of Columbia	107
Florida	369
Georgia	2,433
Idaho	23
Illinois	4,170
Indiana	4,725
Iowa	2,314
Kansas	1,083
Kentucky	3,513
Louisiana	1,053
Maine	1,325
Maryland	1,319
Massachusetts	2,031
Michigan	2,181
Minnesota	729
Mississippi	1,579
Missouri	3,372
Montana	15
Nebraska	356
Nevada	18
New Hampshire	703
New Jersey	1,056
New Mexico	122
New York	6,084
North Carolina	3,142
Ohio	6,460
Oregon	181
Pennsylvania	6,497
Rhode Island	234
South Carolina	1,588

Table V. (Continued).

Tennessee	3,533
Texas	2,276
Utah	148
Vermont	803
Virginia	2,794
Washington	47
West Virginia	1,367
Wisconsin	1,785
Wyoming	2
Males, 45,309; Females, 31,586	76,895

Total number of Idiotic Chinese was 5, and of Indians 84, in 1880.

Native Idiots	72,888	White Idiots	67,316
Foreign do.	4,007	Coloured do.	9,579
	76,895		76,895

Location of Idiots.

In Hospitals and Asylums	1,141
Training Schools	2,429*
Almshouses	5,837
Benevolent Institutions	241
Jails	47
At Home	67,200
	76,895

* About 3 per cent.

APPENDIX F.

TABLE VI.—Showing the Number of Idiots in Training Schools for the Feeble-minded in 1880.*

TRAINING SCHOOLS.	LOCATION.	No.
Connecticut.		
Connecticut Schools for Imbeciles	Lakeville	81
Illinois.		
*Illinois Asylum for Feeble-minded Children	Illinois	306
Indiana.		
*Indiana Asylum for Feeble-minded Children	Knightstown	49
Iowa.		
*Iowa State Asylum for Feeble-minded Children	Glenwood	159
Hospital for Idiots, Imbeciles, &c.	Davenport	8
Kentucky.		
Institution for the Education and Training of Feeble-minded Children	Frankfort	131
Massachusetts.		
Massachusetts School for Idiotic and Feeble-minded Youth	South Boston	113
Private Institution for Education of Feeble-minded Youth	Barre	78
Hillside School for Backward and Feeble-minded Children	Fayfille	11
Minnesota.		
†Minnesota Institution for Feeble-minded Children	Faribault	19
New York.		
New York Asylum for Idiots	Syracuse	295
Newark ,, ,, (Custodial Branch)	Newark	95
Idiot Asylum, Randall's Island	New York	212
Ohio.		
†Ohio State Institution for the Education of Imbecile Youth	Columbus	549
Pennsylvania.		
Pennsylvanian Training School for Feeble-minded Children	Media	323
Total (Males 1,390, Females 1,039)		2429

* Extracted from the Compendium of the "Tenth Census of the United States" In 2 parts, 1880. Part 2, p. 1686.
† State Training Schools.

APPENDIX G.

In addition to the information already afforded in regard to the expenses incurred in the care and treatment of the insane in the United States, the following particulars may be of interest to the reader. I have obtained them from the paper of Dr. Dana, of New York, already referred to in this work:—

In 1882 there was about £8,000,000 invested in institutions for the insane, at an average cost of over £100,000 each. Taking eighty asylums, *the average cost of construction* (generally exclusive of land) was £250 per bed. It takes about £2,160,000 a year "to run" hospitals for the insane, or £16,400 for 131 institutions, exclusive of interest. With interest, the total annual expenditure for the care of the insane amounts to £2,400,000.

The average annual cost per patient for fifty-five hospitals in 1878, as given by Dr. J. A. Reed, was £40; by Dr. Hawthorne for eighty hospitals in the same year, £49. At the State Asylum, Worcester, Mass., it was £41; at the New York State Asylum, Utica, £56; at the State Asylum, Northampton, £36; at the Pennsylvania State Asylum, Harrisburg, £46; at the Willard Asylum, N.Y., £32; at the New Hampshire Asylum, Concord, £52; at the Maryland Hospital for the Insane, Baltimore, £43; at the Brattleboro Asylum, Vermont, £36; at the State Hospital for the Insane, Danvers, £36; at the Government Hospital for the Insane, Washington, £44; at the Connecticut Hospital for the Insane, Middletown, £39; at the Wisconsin State Hospital, Mendota, £47.

November, 1884.

CATALOGUE OF WORKS
PUBLISHED BY
H. K. LEWIS
136 GOWER STREET, LONDON, W.C.

G. GRANVILLE BANTOCK, M.D., F.R.C.S. EDIN.
Surgeon to the Samaritan Free Hospital for Women and Children.

I.
ON THE USE AND ABUSE OF PESSARIES. Second Edition, with Illustrations, 8vo, 5s.

II.
A PLEA FOR EARLY OVARIOTOMY. Demy 8vo, 2s.

FANCOURT BARNES, M.D., M.R.C.P.
Physician to the Chelsea Hospital for Women; Obstetric Physician to the Great Northern Hospital, &c.

A GERMAN-ENGLISH DICTIONARY OF WORDS AND TERMS USED IN MEDICINE AND ITS COGNATE SCIENCES. Square 12mo, Roxburgh binding, 9s.

ASHLEY W. BARRETT, M.B. LOND., M.R.C.S., L.D.S.
Dental Surgeon to the London Hospital, &c.

DENTAL SURGERY FOR GENERAL PRACTITIONERS AND STUDENTS OF MEDICINE. With Illustrations, crown 8vo. [*In the press.*
[Lewis's Practical Series].

ROBERTS BARTHOLOW, M.A., M.D., LL.D.
Professor of Materia Medica and Therapeutics, in the Jefferson Medical College of Philadelphia, &c., &c.

I.
A TREATISE ON THE PRACTICE OF MEDICINE, FOR THE USE OF STUDENTS AND PRACTITIONERS. Fifth Edition, with Illustrations, large 8vo, 21s. [*Just published.*

II.
A PRACTICAL TREATISE ON MATERIA MEDICA AND THERAPEUTICS. Fifth Edition, Revised and Enlarged, 8vo, 18s.
[*Just published.*

GEO. M. BEARD, A.M., M.D.
Fellow of the New York Academy of Medicine; Member of the American Academy of Medicine, &c.
AND
A. D. ROCKWELL, A.M., M.D.
Fellow of the New York Academy of Medicine; Member of the American Academy of Medicine, &c.

A PRACTICAL TREATISE ON THE MEDICAL AND SURGICAL USES OF ELECTRICITY. Including Localized and General Faradization; Localized and Central Galvanization; Franklinization; Electrolysis and Galvano-Cautery. Fourth Edition. With nearly 200 Illustrations, roy. 8vo, 28s. [*Just published.*

A. HUGHES BENNETT, M.D.
Member of the Royal College of Physicians of London; Physician to the Hospital for Epilepsy and Paralysis, Regent's Park, and Assistant Physician to the Westminster Hospital.

I.
A PRACTICAL TREATISE ON ELECTRO-DIAGNOSIS IN DISEASES OF THE NERVOUS-SYSTEM. With Illustrations, 8vo, 8s. 6d.

II.
ILLUSTRATIONS OF THE SUPERFICIAL NERVES AND MUSCLES, WITH THEIR MOTOR POINTS, A knowledge of which is essential in the Art of Electro-Diagnosis. (Extracted from the above). 8vo, paper cover 1s. 6d., cloth 2s.

III.
ON EPILEPSY: ITS NATURE AND TREATMENT. 8vo, 2s. 6d.

DR. THEODOR BILLROTH.
Professor of Surgery in Vienna.

GENERAL SURGICAL PATHOLOGY AND THERAPEUTICS. In Fifty-one Lectures. A Text-book for Students and Physicians. With additions by Dr. ALEXANDER VON WINIWARTER, Professor of Surgery in Luttich. Translated from the Fourth German edition with the special permission of the Author, and revised from the Tenth edition, by C. E. HACKLEY, A.M., M.D. Copiously illustrated, 8vo, 18s.

G. H. BRANDT, M.D.

I.
ROYAT (LES BAINS) IN AUVERGNE, ITS MINERAL WATERS AND CLIMATE. With Frontispiece and Map. Second edition, crown 8vo, 2s. 6d.

II.
HAMMAM R'IRHA, ALGIERS. A Winter Health Resort and Mineral Water Cure Combined. With Frontispiece and Map, crown 8vo, 2s. 6d.

GURDON BUCK, M.D.

CONTRIBUTIONS TO REPARATIVE SURGERY; SHOWing its Application to the Treatment of Deformities, produced by Destructive Disease or Injury; Congenital Defects from Arrest or Excess of Development; and Cicatricial Contractions from Burns. Illustrated by numerous Engravings, large 8vo, 9s.

ALFRED H. CARTER, M.D. LOND.
Member of the Royal College of Physicians; Physician to the Queen's Hospital, Birmingham; Examiner in Medicine for the University of Aberdeen, &c.

ELEMENTS OF PRACTICAL MEDICINE. Third Edition, crown 8vo, 9s. [*Now ready.*

P. CAZEAUX.
Adjunct Professor in the Faculty of Medicine of Paris, &c.
AND
S. TARNIER.
Professor of Obstetrics and Diseases of Women and Children in the Faculty of Medicine of Paris.

OBSTETRICS: THE THEORY AND PRACTICE; including the Diseases of Pregnancy and Parturition, Obstetrical Operations, &c. Seventh Edition, edited and revised by ROBERT J. HESS, M.D., with twelve full-page plates, five being coloured, and 165 wood-engravings, 1081 pages, roy. 8vo, 35s. [*Just ready.*

JOHN COCKLE, M.A., M.D.
Physician to the Royal Free Hospital.

ON INTRA-THORACIC CANCER. 8vo, 4s. 6d.

W. H. CORFIELD, M.A., M.D. OXON.
Professor of Hygiene and Public Health in University College, London.

DWELLING HOUSES: Their Sanitary Construction and Arrangements. Second Edition, with Illustrations. [*In the press.*

J. THOMPSON DICKSON, M.A., M.B. CANTAB.
Late Lecturer on Mental Diseases at Guy's Hospital.

THE SCIENCE AND PRACTICE OF MEDICINE IN RELATION TO MIND, the Pathology of the Nerve Centres, and the Jurisprudence of Insanity, being a course of Lectures delivered at Guy's Hospital. Illustrated by Chromo-lithographic Drawings and Physiological Portraits. 8vo, 14s.

HORACE DOBELL, M.D.
Consulting Physician to the Royal Hospital for Diseases of the Chest, &c.

I.
ON DIET AND REGIMEN IN SICKNESS AND HEALTH, and on the Interdependence and Prevention of Diseases and the Diminution of their Fatality. Seventh Edition, 8vo, 10s. 6d.

II.
AFFECTIONS OF THE HEART AND IN ITS NEIGHBOURHOOD. Cases, Aphorisms, and Commentaries. Illustrated by the heliotype process. 8vo, 6s 6d.

JOHN EAGLE.
Member of the Pharmaceutical Society.

A NOTE-BOOK OF SOLUBILITIES. Arranged chiefly for the use of Prescribers and Dispensers. 12mo, 2s. 6d.

JOHN ERIC ERICHSEN.
Holme Professor of Clinical Surgery in University College; Senior Surgeon to University College Hospital, &c.

MODERN SURGERY; Its Progress and Tendencies. Being the Introductory Address delivered at University College at the opening of the Session 1873-74. Demy 8vo, 1s.

DR. FERBER.

MODEL DIAGRAM OF THE ORGANS IN THE THORAX AND UPPER PART OF THE ABDOMEN. With Letter-press Description. In 4to, coloured, 5s.

AUSTIN FLINT, Jr., M.D.
Professor of Physiology and Physiological Anatomy in the Bellevue Medical College, New York; attending Physician to the Bellevue Hospital, &c.

I.
A TEXT-BOOK OF HUMAN PHYSIOLOGY; DESIGNED for the Use of Practitioners and Students of Medicine. New edition, Illustrated by plates, and 313 wood engravings, large 8vo, 28s.

II.
THE PHYSIOLOGY OF THE SPECIAL SENSES AND GENERATION; (Being Vol. V of the Physiology of Man). Roy. 8vo, 18s.

J. MILNER FOTHERGILL, M.D.
Member of the Royal College of Physicians of London; Physician to the City of London Hospital for Diseases of the Chest, Victoria Park, &c.

I.
THE HEART AND ITS DISEASES, WITH THEIR TREATMENT; INCLUDING THE GOUTY HEART. Second Edition, entirely re-written, copiously illustrated with woodcuts and lithographic plates. 8vo. 16s.

II.
INDIGESTION, BILIOUSNESS, AND GOUT IN ITS PROTEAN ASPECTS.
PART I.—INDIGESTION AND BILIOUSNESS. Post 8vo, 7s. 6d.
PART II.—GOUT IN ITS PROTEAN ASPECTS. Post 8vo, 7s. 6d.

III.
HEART STARVATION. (Reprinted from the Edinburgh Medical Journal), 8vo, 1s.

ERNEST FRANCIS, F.C.S.
Demonstrator of Practical Chemistry, Charing Cross Hospital.

PRACTICAL EXAMPLES IN QUANTITATIVE ANALYSIS, forming a Concise Guide to the Analysis of Water, &c. Illustrated, fcap. 8vo, 2s. 6d.

HENEAGE GIBBES, M.D.
Lecturer on Physiology and Histology in the Medical School of Westminster Hospital; late Curator of the Anatomical Museum at King's College.

PRACTICAL HISTOLOGY AND PATHOLOGY. Second Edit. revised and enlarged. Crown 8vo, 5s.

C. A. GORDON, M.D., C.B.
Deputy Inspector General of Hospitals, Army Medical Department.

REMARKS ON ARMY SURGEONS AND THEIR WORKS. Demy 8vo, 5s.

W. R. GOWERS, M.D., F.R.C.P. M.R.C.S.
Physician to University College Hospital, &c.

DIAGRAMS FOR THE RECORD OF PHYSICAL SIGNS. In books of 12 sets of figures, 1s. Ditto, unbound, 1s.

SAMUEL D. GROSS, M.D., LL.D., D.C.L., OXON.
Professor of Surgery in the Jefferson Medical College of Philadelphia.

A PRACTICAL TREATISE ON THE DISEASES, INJURIES, AND MALFORMATIONS OF THE URINARY BLADDER, THE PROSTATE GLAND; AND THE URETHRA. Third Edition, revised and edited by S. W. GROSS, A.M., M.D., Surgeon to the Philadelphia Hospital. Illustrated by 170 engravings, 8vo, 18s.

SAMUEL W. GROSS, A.M., M.D.
Surgeon to, and Lecturer on Clinical Surgery in, the Jefferson Medical College Hospital, and the Philadelphia Hospital, &c.

A PRACTICAL TREATISE ON TUMOURS OF THE MAMMARY GLAND: embracing their Histology, Pathology, Diagnosis, and Treatment. With Illustrations, 8vo, 10s. 6d.

WILLIAM A. HAMMOND, M.D.
Professor of Mental and Nervous Diseases in the Medical Department of the University of the City of New York, &c.

I.
A TREATISE ON THE DISEASES OF THE NERVOUS SYSTEM. Seventh edition, with 112 Illustrations, large 8vo, 25s.

II.
A TREATISE ON INSANITY. Large 8vo, 25s.
[*Just published.*]

III.
SPIRITUALISM AND ALLIED CAUSES AND CONDITIONS OF NERVOUS DERANGEMENT. With Illustrations, post 8vo, 8s. 6d.

ALEXANDER HARVEY, M.A., M.D.
Emeritus Professor of Materia Medica in the University of Aberdeen; Consulting Physician to the Aberdeen Royal Infirmary, &c.

FIRST LINES OF THERAPEUTICS; as based on the Modes and the Processes of Healing, as occurring Spontaneously in Disease; and on the Modes and the Processes of Dying, as resulting Naturally from Disease. In a series of Lectures. Post 8vo, 5s.

ALEXANDER HARVEY, M.D.
Emeritus Professor of Materia Medica in the University of Aberdeen, &c.
AND
ALEXANDER DYCE DAVIDSON, M.D.
Professor of Materia Medica in the University of Aberdeen.

SYLLABUS OF MATERIA MEDICA FOR THE USE OF TEACHERS AND STUDENTS. Based on a selection or definition of subjects in teaching and examining; and also on an estimate of the relative values of articles and preparations in the British Pharmacopœia with doses affixed. Seventh Edition, 32mo.
[*In preparation.*]

GRAILY HEWITT, M.D.
Professor of Midwifery and Diseases of Women in University College, Obstetrical Physician to University College Hospital, &c.

OUTLINES OF PICTORIAL DIAGNOSIS OF DISEASES OF WOMEN. Fol. 6s.

BERKELEY HILL, M.B. LOND., F.R.C.S.
Professor of Clinical Surgery in University College; Surgeon to University College Hospital and to the Lock Hospital.

THE ESSENTIALS OF BANDAGING. For Managing Fractures and Dislocations; for administering Ether and Chloroform; and for using other Surgical Apparatus. Fifth Edition, revised and much enlarged, with Illustrations, fcap. 8vo, 5s.

BERKELEY HILL, M.B. LOND., F.R.C.S.
Professor of Clinical Surgery in University College; Surgeon to University College Hospital and to the Lock Hospital.

AND

ARTHUR COOPER, L.R.C.P., M.R.C.S.
Late House Surgeon to the Lock Hospital, &c.

I.

SYPHILIS AND LOCAL CONTAGIOUS DISORDERS. Second Edition, entirely re-written, royal 8vo, 18s.

II.

THE STUDENT'S MANUAL OF VENEREAL DISEASES. Being a Concise Description of those Affections and of their Treatment. Third Edition, post 8vo, 2s. 6d.

HINTS TO CANDIDATES FOR COMMISSIONS IN THE PUBLIC MEDICAL SERVICES, WITH EXAMINATION QUESTIONS, VOCABULARY OF HINDUSTANI MEDICAL TERMS, ETC. 8vo, 2s.

SIR W. JENNER, Bart., M.D.
Physician in Ordinary to H.M. the Queen, and to H.R.H. the Prince of Wales.

THE PRACTICAL MEDICINE OF TO-DAY: Two Addresses delivered before the British Medical Association, and the Epidemiological Society, (1869). Small 8vo, 1s. 6d.

C. M. JESSOP, M.R.C P.
Associate of King's College, London; Brigade Surgeon H.M. British Forces.

ASIATIC CHOLERA, being a Report on an Outbreak of Epidemic Cholera in 1876 at a Camp near Murree in India. With map, demy 8vo, 2s. 6d.

GEORGE LINDSAY JOHNSON, M.A., M.B., B.C. CANTAB.
Clinical Assistant, late House Surgeon and Chloroformist, Royal Westminster Ophthalmic Hospital; Medical and Surgical Registrar, &c.

A NEW METHOD OF TREATING CHRONIC GLAU-COMA, based on Recent Researches into its Pathology. With Illustrations and coloured frontispiece, demy 8vo, 3s. 6d.

NORMAN W. KINGSLEY, M.D.S., D.D.S.
President of the Board of Censors of the State of New York; Member of the American Academy of Dental Science, &c.

A TREATISE ON ORAL DEFORMITIES AS A BRANCH OF MECHANICAL SURGERY. With over 350 Illustrations, 8vo, 16s.

E. A. KIRBY, M.D., M.R.C.S. ENG.
Late Physician to the City Dispensary.

I.
A PHARMACOPŒIA OF SELECTED REMEDIES, WITH THERAPEUTIC ANNOTATIONS, Notes on Alimentation in Disease, Air, Massage, Electricity and other Supplementary Remedial Agents, and a Clinical Index; arranged as a Handbook for Prescribers. Sixth Edition, enlarged and revised, demy 4to, 7s.

II.
ON THE VALUE OF PHOSPHORUS AS A REMEDY FOR LOSS OF NERVE POWER. Fifth Edition, 8vo, 2s. 6d.

J. WICKHAM LEGG, F.R.C.P.
Assistant Physician to Saint Bartholomew's Hospital, and Lecturer on Pathological Anatomy in the Medical School.

I.
ON THE BILE, JAUNDICE, AND BILIOUS DISEASES. With Illustrations in chromo-lithography, 719 pages, roy. 8vo, 25s.

II.
A GUIDE TO THE EXAMINATION OF THE URINE; intended chiefly for Clinical Clerks and Students. Fifth Edition, revised and enlarged, with additional Illustrations, fcap. 8vo, 2s. 6d.

III.
A TREATISE ON HÆMOPHILIA, SOMETIMES CALLED THE HEREDITARY HÆMORRHAGIC DIATHESIS. Fcap. 4to, 7s. 6d.

DR. GEORGE LEWIN.
Professor at the Fr. Wilh. University, and Surgeon-in-Chief of the Syphilitic Wards and Skin Disease Wards of the Charité Hospital, Berlin.

THE TREATMENT OF SYPHILIS WITH SUBCUTA-NEOUS SUBLIMATE INJECTIONS. Translated by DR. CARL PRŒGLE, and DR. E. H. GALE, late Surgeon United States Army. Small 8vo, 7s.

LEWIS'S PRACTICAL SERIES.

Under this title Mr. LEWIS purposes publishing a complete Series of Monographs, embracing the various branches of Medicine and Surgery.

The volumes, written by well-known Hospital Physicians and Surgeons recognized as authorities in the subjects of which they treat, are in active preparation. The works are intended to be of a THOROUGHLY PRACTICAL nature, calculated to meet the requirements of the general practitioner, and to present the most recent information in a compact and readable form; the volumes will be handsomely got up, and issued at low prices, varying with the size of the works.

BODILY DEFORMITIES AND THEIR TREATMENT: A HANDBOOK OF PRACTICAL ORTHOPÆDICS. By H. A. REEVES, F.R.C.S. Edin., Senior Assistant Surgeon and Teacher of Practical Surgery at the London Hospital; Surgeon to the Royal Orthopædic Hospital, &c. With numerous Illustrations, cr. 8vo. [*In the press.*

DENTAL SURGERY FOR GENERAL PRACTITIONERS AND STUDENTS OF MEDICINE. By ASHLEY W. BARRETT, M.B. Lond., M.R.C.S., L.D.S., Dental Surgeon to, and Lecturer on Dental Surgery and Pathology in the Medical School of, the London Hospital. With Illustrations, crown 8vo. [*In the press.*

*** Further volumes will be announced in due course.

LEWIS'S POCKET MEDICAL VOCABULARY.
[*In the Press.*

J. S. LOMBARD, M.D.
Formerly Assistant Professor of Physiology in Harvard College.

I.
EXPERIMENTAL RESEARCHES ON THE REGIONAL TEMPERATURE OF THE HEAD, under Conditions of Rest, Intellectual Activity and Emotion. With Illustrations, 8vo, 8s.

II.
ON THE NORMAL TEMPERATURE OF THE HEAD. 8vo, 5s.

WILLIAM THOMPSON LUSK, A.M., M.D.
Professor of Obstetrics and Diseases of Women in the Bellevue Hospital Medical College, &c.
THE SCIENCE AND ART OF MIDWIFERY. Second Edition, with numerous Illustrations, 8vo, 18s.

JOHN MACPHERSON, M.D.
Inspector-General of Hospitals H.M. Bengal Army (Retired).
Author of "Cholera in its Home," &c.
ANNALS OF CHOLERA FROM THE EARLIEST PERIODS TO THE YEAR 1817. With a map. Demy 8vo, 7s. 6d.

DR. V. MAGNAN.
Physician to St. Anne Asylum, Paris; Laureate of the Institute.
ON ALCOHOLISM, the Various Forms of Alcoholic Delirium and their Treatment. Translated by W. S. GREENFIELD, M.D., M.R.C.P. 8vo, 7s. 6d.

A. COWLEY MALLEY, B.A., M.B., B.CH. T.C.D.

PHOTO-MICROGRAPHY; including a description of the Wet Collodion and Gelatino-Bromide Processes, together with the best methods of Mounting and Preparing Microscopic Objects for Photo-Micrography. Second Edition, with Photographs and Illustrations, crown 8vo. [*In the press.*

PATRICK MANSON, M.D., C.M.
Amoy, China.

THE FILARIA SANGUINIS HOMINIS; AND CERTAIN NEW FORMS OF PARASITIC DISEASE IN INDIA, CHINA, AND WARM COUNTRIES. Illustrated with Plates and Charts. 8vo, 10s. 6d.

PROFESSOR MARTIN.

MARTIN'S ATLAS OF OBSTETRICS AND GYNÆCOLOGY. Edited by A. MARTIN, Docent in the University of Berlin. Translated and edited with additions by FANCOURT BARNES, M.D., M.R.C.P., Physician to the Chelsea Hospital for Women; Obstetric Physician to the Great Northern Hospital; and to the Royal Maternity Charity of London, &c. Medium 4to, Morocco half bound, 31s. 6d. net.

WILLIAM MARTINDALE, F.C.S.
Late Examiner of the Pharmaceutical Society, and late Teacher of Pharmacy and Demonstrator of Materia Medica at University College.
AND
W. WYNN WESTCOTT, M.B. LOND.
Deputy Coroner for Central Middlesex.

THE EXTRA PHARMACOPŒIA of Unofficial Drugs and Chemical and Pharmaceutical Preparations, with References to their Use abstracted from the Medical Journals and a Therapeutic Index of Diseases and Symptoms. Third Edition, revised with numerous additions, limp roan, med. 24mo, 7s., and an edition in fcap. 8vo, with room for marginal notes, cloth, 7s. [*Now ready.*

J. F. MEIGS, M.D.
Consulting Physician to the Children's Hospital, Philadelphia.
AND
W. PEPPER, M.D.
Lecturer on Clinical Medicine in the University of Pennsylvania.

A PRACTICAL TREATISE ON THE DISEASES OF CHILDREN. Seventh Edition, revised and enlarged, roy. 8vo, 28s.

Wm. JULIUS MICKLE, M.D., M.R.C.P. LOND.
Member of the Medico-Psychological Association of Great Britain and Ireland; Member of the Clinical Society, London; Medical Superintendent, Grove Hall Asylum, London.

GENERAL PARALYSIS OF THE INSANE. 8vo, 10s.

KENNETH W. MILLICAN, B.A. CANTAB., M.R.C.S.

THE EVOLUTION OF MORBID GERMS: A Contribution to Transcendental Pathology. Cr. 8vo, 3s. 6d.

E. A. MORSHEAD, M.R.C.S., L.R.C.P.
Assistant to the Professor of Medicine in University College, London.

TABLES OF THE PHYSIOLOGICAL ACTION OF DRUGS. Fcap. 8vo, 1s.

A. STANFORD MORTON, M.B., F.R.C.S. ED.
Senior Assistant Surgeon, Royal South London Ophthalmic Hospital.

REFRACTION OF THE EYE: Its Diagnosis, and the Correction of its Errors, with Chapter on Keratoscopy. Second edit., with Illustrations, small 8vo, 2s. 6d.

WILLIAM MURRELL, M.D., F.R.C.P.
Lecturer on Materia Medica and Therapeutics at Westminster Hospital; Examiner in Materia Medica, University of Edinburgh.

I.
WHAT TO DO IN CASES OF POISONING. Fourth Edition, revised and enlarged, royal 32mo, 3s. 6d.

II.
NITRO-GLYCERINE AS A REMEDY FOR ANGINA PECTORIS. Crown 8vo, 3s. 6d.

WILLIAM NEWMAN, M.D. LOND., F.R.C.S.
Surgeon to the Stamford Infirmary.

SURGICAL CASES: Mainly from the Wards of the Stamford, Rutland, and General Infirmary, 8vo, paper boards, 4s. 6d.

DR. FELIX von NIEMEYER.
Late Professor of Pathology and Therapeutics; Director of the Medical Clinic of the University of Tübingen.

A TEXT-BOOK OF PRACTICAL MEDICINE, WITH PARTICULAR REFERENCE TO PHYSIOLOGY AND PATHOLOGICAL ANATOMY. Translated from the Eighth German Edition, by special permission of the Author, by GEORGE H. HUMPHREY, M.D., and CHARLES E. HACKLEY, M.D., Revised Edition, 2 vols., large 8vo, 36s.

C. F. OLDHAM, M.R.C.S., L.R.C.P.

Surgeon H.M. Indian Forces; late in Medical charge of the Dalhousie Sanitarium.

WHAT IS MALARIA? and why is it most intense in hot climates? An explanation of the Nature and Cause of the so-called Marsh Poison, with the Principles to be observed for the Preservation of Health in Tropical Climates and Malarious Districts. Demy 8vo, 7s. 6d.

G. OLIVER, M.D., M.R.C.P.

I.

THE HARROGATE WATERS: Data Chemical and Therapeutical, with notes on the Climate of Harrogate. Addressed to the Medical Profession. Crown 8vo, with Map of the Wells, 3s. 6d.

II.

ON BEDSIDE URINE TESTING: including Quantitative Albumen and Sugar. Third edition, revised and enlarged, fcap. 8vo.

[*In the press.*

JOHN S. PARRY, M.D.

Obstetrician to the Philadelphia Hospital, Vice-President of the Obstetrical and Pathological Societies of Philadelphia, &c.

EXTRA-UTERINE PREGNANCY; Its Causes, Species, Pathological Anatomy, Clinical History, Diagnosis, Prognosis and Treatment. 8vo, 8s.

E. RANDOLPH PEASLEE, M.D., LL.D.

Late Professor of Gynæcology in the Medical Department of Dartmouth College; President of the New York Academy of Medicine, &c., &c.

OVARIAN TUMOURS: Their Pathology, Diagnosis, and Treatment, especially by Ovariotomy. Illustrations, roy. 8vo, 16s.

G. V. POORE, M.D., F.R.C.P.

Professor of Medical Jurisprudence, University College; Assistant Physician to, and Physician in charge of the Throat Department of University College Hospital.

LECTURES ON THE PHYSICAL EXAMINATION OF THE MOUTH AND THROAT. With an Appendix of Cases. 8vo, 3s. 6d.

R. DOUGLAS POWELL, M.D., F.R.C.P. LOND.

Physician to the Middlesex Hospital, and Physician to the Hospital for Consumption and Diseases of the Chest at Brompton.

DISEASES OF THE LUNGS AND PLEURÆ. Third Edition, rewritten and enlarged. With Illustrations, 8vo.

[*In preparation.*

AMBROSE L. RANNEY, A.M., M.D.
Adjunct Professor of Anatomy in the University of New York, etc.

THE APPLIED ANATOMY OF THE NERVOUS SYSTEM, being a study of this portion of the Human Body from a standpoint of its general interest and practical utility, designed for use as a Text-book and a Work of Reference. With 179 Illustrations, 8vo, 20s.

H. A. REEVES, F.R.C.S. ED.
Senior Assistant Surgeon and Teacher of Practical Surgery at the London Hospital; Surgeon to the Royal Orthopædic Hospital, &c.

BODILY DEFORMITIES AND THEIR TREATMENT: A Handbook of Practical Orthopædics. With numerous Illustrations, crown 8vo. [*In the press.*
Lewis's Practical Series].

RALPH RICHARDSON, M.A., M.D.
Fellow of the College of Physicians, Edinburgh.

ON THE NATURE OF LIFE: An Introductory Chapter to Pathology. Second Edition, revised and enlarged. Fcap. 4to, 10s. 6d.

W. RICHARDSON, M.A., M.D., M.R.C.P.

REMARKS ON DIABETES, ESPECIALLY IN REFERENCE TO TREATMENT. Demy 8vo, 4s. 6d.

SYDNEY RINGER, M.D.
Professor of the Principles and Practice of Medicine in University College; Physician to, and Professor of Clinical Medicine in, University College Hospital.

I.
A HANDBOOK OF THERAPEUTICS. Tenth Edition, 8vo, 15s.

II.
ON THE TEMPERATURE OF THE BODY AS A MEANS OF DIAGNOSIS AND PROGNOSIS IN PHTHISIS. Second Edition, small 8vo, 2s. 6d.

FREDERICK T. ROBERTS, M.D., B.SC., F.R.C.P.
Examiner in Medicine at the Royal College of Surgeons; Professor of Therapeutics in University College; Physician to University College Hospital; Physician to Brompton Consumption Hospital, &c.

I.
A HANDBOOK OF THE THEORY AND PRACTICE OF MEDICINE. Fifth Edition, with Illustrations, in one volume, large 8vo, 21s.

II.
NOTES ON MATERIA MEDICA AND PHARMACY. Fcap. 8vo, 7s. 6d. [*Now ready.*

D. B. St. JOHN ROOSA, M.A., M.D.

Professor of Diseases of the Eye and Ear in the University of the City of New York; Surgeon to the Manhattan Eye and Ear Hospital; Consulting Surgeon to the Brooklyn Eye and Ear Hospital, &c., &c.

A PRACTICAL TREATISE ON THE DISEASES OF THE EAR, including the Anatomy of the Organ. Fourth Edition, Illustrated by wood engravings and chromo-lithographs, large 8vo, 22s.

J. BURDON SANDERSON, M.D., LL.D., F.R.S.

Jodrell Professor of Physiology in University College, London.

UNIVERSITY COLLEGE COURSE OF PRACTICAL EXERCISES IN PHYSIOLOGY. With the co-operation of F. J. M. PAGE, B.Sc., F.C.S.; W. NORTH, B.A., F.C.S., and AUG. WALLER, M.D. Demy 8vo, 3s. 6d.

W. H. O. SANKEY, M.D. LOND., F.R.C.P.

Late Lecturer on Mental Diseases, University College and School of Medicine for Women, London; Formerly Medical Superintendent (Female Department) of Hanwell Asylum; President of Medico-Psychological Society, &c.

LECTURES ON MENTAL DISEASE. Second Edition, with coloured plates, 8vo, 12s. 6d. [*Now ready.*

ALDER SMITH, M.B. LOND., F.R.C.S.

Resident Medical Officer, Christ's Hospital, London.

RINGWORM: Its Diagnosis and Treatment. Second Edition, rewritten and enlarged. With Illustrations, fcap. 8vo, 4s. 6d.

J. LEWIS SMITH, M.D.

Physician to the New York Infants' Hospital; Clinical Lecturer on Diseases of Children in Bellevue Hospital Medical College.

A TREATISE ON THE DISEASES OF INFANCY AND CHILDHOOD. Fifth Edition, with Illustrations, large 8vo, 21s.

FRANCIS W. SMITH, M.B., B.S.

THE LEAMINGTON WATERS; CHEMICALLY, THERAPEUTICALLY AND CLINICALLY CONSIDERED; with observations on the climate of Leamington. With Illustrations, crown 8vo, 2s. 6d.

JAMES STARTIN, M.B., M.R.C.S.

Surgeon and Joint Lecturer to St. John's Hospital for Diseases of the Skin.

LECTURES ON THE PARASITIC DISEASES OF THE SKIN. VEGETOID AND ANIMAL. With Illustrations, Crown 8vo, 3s. 6d.

HENRY R. SWANZY, M.A., M.B., F.R.C.S.I.

Examiner in Ophthalmic Surgery, University of Dublin; Surgeon to the National Eye and Ear Infirmary, Dublin; Ophthalmic Surgeon at the Adelaide Hospital, Dublin.

HANDBOOK OF DISEASES OF THE EYE AND THEIR TREATMENT. Illustrated with wood-engravings, colour tests, etc., large post 8vo, 10s. 6d. [*Now ready.*

LEWIS A. STIMSON, B.A., M.D.
Surgeon to the Presbyterian Hospital; Professor of Pathological Anatomy in the Medical Faculty of the University of the City of New York.

A MANUAL OF OPERATIVE SURGERY. With three hundred and thirty-two Illustrations. Post 8vo, 10s. 6d.

HUGH OWEN THOMAS, M.R.C.S.

I.

DISEASES OF THE HIP, KNEE, AND ANKLE JOINTS, with their Deformities, treated by a new and efficient method. With an Introduction by RUSHTON PARKER, F.R.C.S, Lecturer on Surgery at the School of Medicine, Liverpool. Second Edition, 8vo, 25s.

II.

CONTRIBUTIONS TO MEDICINE AND SURGERY:—

PART 1.—Intestinal Obstruction; with an Appendix on the Action of Remedies. 10s.
,, 2.—The Principles of the Treatment of Joint Disease, Inflammation, Anchylosis, Reduction of Joint Deformity, Bone Setting. 5s.
,, 5.—On Fractures of the Lower Jaw. 1s.
,, 8.—The Inhibition of Nerves by Drugs. Proof that Inhibitory Nerve-Fibres do not exist. 1s.

(Parts 3, 4, 6, 7, 9, 10, are expected shortly).

J. ASHBURTON THOMPSON, M.R.C.S.
Late Surgeon at King's Cross to the Great Northern Railway Company.

FREE PHOSPHORUS IN MEDICINE WITH SPECIAL REFERENCE TO ITS USE IN NEURALGIA. A contribution to Materia Medica and Therapeutics. An account of the History, Pharmaceutical Preparations, Dose, Internal Administration, and Therapeutic uses of Phosphorus; with a Complete Bibliography of this subject, referring to nearly 200 works upon it. Demy 8vo, 7s. 6d.

J. C. THOROWGOOD, M.D.
Assistant Physician to the City of London Hospital for Diseases of the Chest.

THE CLIMATIC TREATMENT OF CONSUMPTION AND CHRONIC LUNG DISEASES. Third Edition, post 8vo, 3s 6d.

EDWARD T. TIBBITS, M.D. LOND.
Physician to the Bradford Infirmary; and to the Bradford Fever Hospital.

MEDICAL FASHIONS IN THE NINETEENTH CENTURY, including a Sketch of Bacterio-Mania and the Battle of the Bacilli. Crown 8vo, 2s. 6d.

LAURENCE TURNBULL, M.D., PH.G.
Aural Surgeon to Jefferson Medical College Hospital, &c., &c.

ARTIFICIAL ANÆSTHESIA: A Manual of Anæsthetic Agents, and their Employment in the Treatment of Disease. Second Edition, with Illustrations, crown 8vo, 6s.

W. H. VAN BUREN, M.D., LL.D.
Professor of Surgery in the Bellevue Hospital Medical College.

DISEASES OF THE RECTUM: And the Surgery of the Lower Bowel. Second Edition, with Illustrations, 8vo, 14s.

RUDOLPH VIRCHOW, M.D.
Professor in the University, and Member of the Academy of Sciences of Berlin, &c., &c.

INFECTION - DISEASES IN THE ARMY, Chiefly Wound Fever, Typhoid, Dysentery, and Diphtheria. Translated from the German by JOHN JAMES, M.B., F.R.C.S. Fcap. 8vo, 1s. 6d.

ALFRED VOGEL, M.D.
Professor of Clinical Medicine in the University of Dorpat, Russia.

A PRACTICAL TREATISE ON THE DISEASES OF CHILDREN. Translated and Edited by H. RAPHAEL, M.D. From the Fourth German Edition, illustrated by six lithographic plates, part coloured, large 8vo, 18s.

A. DUNBAR WALKER, M.D., C.M.

THE PARENT'S MEDICAL NOTE BOOK. Oblong post 8vo, cloth, 1s. 6d.

W. SPENCER WATSON, F.R.C.S. ENG., B.M. LOND.
Surgeon to the Great Northern Hospital; Surgeon to the Royal South London Ophthalmic Hospital.

I.
DISEASES OF THE NOSE AND ITS ACCESSORY CAVITIES. Profusely Illustrated. Demy 8vo, 18s.

II.
EYEBALL-TENSION: Its Effects on the Sight and its Treatment. With woodcuts, p. 8vo, 2s. 6d.

III.
ON ABSCESS AND TUMOURS OF THE ORBIT. Post 8vo, 2s. 6d.

A. DE WATTEVILLE, M.A., M.D., B.SC., M.R.C.S.
Physician in Charge of the Electro-therapeutical Department at St. Mary's Hospital.

A PRACTICAL INTRODUCTION TO MEDICAL ELECTRICITY. Second Edition, re-written and enlarged, copiously Illustrated, 8vo, 9s. [*Just published.*

FRANCIS H. WELCH, F.R.C.S.
Surgeon Major, A.M.D.

ENTERIC FEVER: as Illustrated by Army Data at Home and Abroad, its Prevalence and Modifications, Ætiology, Pathology and Treatment. 8vo, 5s. 6d. [*Just published.*

DR. F. WINCKEL.
Formerly Professor and Director of the Gynæcological Clinic at the University of Rostock.

THE PATHOLOGY AND TREATMENT OF CHILD-BED: A Treatise for Physicians and Students. Translated from the Second German edition, with many additional notes by the Author, by J. R. CHADWICK, M.D., 8vo, 14s.

EDWARD WOAKES, M.D. LOND.
Senior Aural Surgeon and Lecturer on Aural Surgery at the London Hospital; Senior Surgeon to the Hospital for Diseases of the Throat.

ON DEAFNESS, GIDDINESS AND NOISES IN THE HEAD.

Vol. I.—POST-NASAL CATARRH, AND DISEASES OF THE NOSE CAUSING DEAFNESS. With Illustrations, cr. 8vo, 6s. 6d.

Vol. II.—ON DEAFNESS, GIDDINESS AND NOISES IN THE HEAD. Third Edition, with Illustrations, cr. 8vo. [*In preparation.*

E. T. WILSON, B.M. OXON., F.R.C.P. LOND.
Physician to the Cheltenham General Hospital and Dispensary.

DISINFECTANTS AND HOW TO USE THEM.
In Packets of one doz. price 1s.

Clinical Charts For Temperature Observations, etc.

Arranged by W. RIGDEN, M.R.C.S. Price 7s. per 100, or 1s. per dozen.

Each Chart is arranged for four weeks, and is ruled at the back for making notes of cases; they are convenient in size, and are suitable both for hospital and private practice.

PERIODICAL WORKS PUBLISHED BY H. K. LEWIS.

THE NEW SYDENHAM SOCIETY'S PUBLICATIONS. Annual Subscription, One Guinea.
(Report of the Society, with Complete List of Works and other information, gratis on application.)

ARCHIVES OF PEDIATRICS. A Monthly Journal, devoted to the Diseases of Infants and Children. Annual Subscription, 12s. 6d., post free.

THE NEW YORK MEDICAL JOURNAL. A Weekly Review of Medicine. Annual Subscription, One Guinea, post free.

THE THERAPEUTIC GAZETTE. A Monthly Journal, devoted to the Science of Pharmacology, and to the introduction of New Therapeutic Agents. Annual Subscription, 5s., post free.

THE GLASGOW MEDICAL JOURNAL. Published Monthly. Annual Subscription, 20s., post free. Single numbers, 2s. each.

LIVERPOOL MEDICO-CHIRURGICAL JOURNAL, including the Proceedings of the Liverpool Medical Institution. Published twice yearly, 3s. 6d. each.

THE INDIAN MEDICAL JOURNAL. A Journal of Medical and Sanitary Science specially devoted to the Interests of the Medical Services. Annual Subscription, 24s., post free.

THE MIDLAND MEDICAL MISCELLANY AND PROVINCIAL MEDICAL JOURNAL. Annual Subscription, 7s. 6d., post free.

TRANSACTIONS OF THE COLLEGE OF PHYSICIANS OF PHILADELPHIA. Volumes I to VI., now ready, 8vo, 10s. 6d. each.

*** MR. LEWIS has transactions with the leading publishing firms in America for the sale of his publications in that country. Arrangements are made in the interests of Authors either for sending a number of copies of their works to the United States, or having them reprinted there, as may be most advantageous.

Mr. Lewis's publications can be procured of any bookseller in any part of the world.

www.ingramcontent.com/pod-product-compliance
Lightning Source LLC
Chambersburg PA
CBHW031336230426
43670CB00006B/348